Lecture Notes in Mathematics

Edited by A. Dold and B. Eckmann

1272

Moshe S. Livšic
Leonid L. Waksman

Commuting Nonselfadjoint
Operators in Hilbert Space

Two Independent Studies

Springer-Verlag
Berlin Heidelberg New York London Paris Tokyo

Authors

Moshe S. Livšic
Dept. of Mathematics
Beer Sheva
Ben-Gurion University of the Negev, Israel

Leonid L. Waksman
ul. 1 konnoi Armii d. 18/9 kv. 6
344029 Rostov-na-Donu, USSR

Mathematics Subject Classification (1980): 15, 35A07, 46, 47, 47D05, 47H06, 73, 81, 93

ISBN 3-540-18316-7 Springer-Verlag Berlin Heidelberg New York
ISBN 0-387-18316-7 Springer-Verlag New York Berlin Heidelberg

This work is subject to copyright. All rights are reserved, whether the whole or part of the material is concerned, specifically the rights of translation, reprinting, re-use of illustrations, recitation, broadcasting, reproduction on microfilms or in other ways, and storage in data banks. Duplication of this publication or parts thereof is only permitted under the provisions of the German Copyright Law of September 9, 1965, in its version of June 24, 1985, and a copyright fee must always be paid. Violations fall under the prosecution act of the German Copyright Law.

© Springer-Verlag Berlin Heidelberg 1987
Printed in Germany

Printing and binding: Druckhaus Beltz, Hemsbach/Bergstr.
2146/3140-543210

Commuting Nonselfadjoint Operators

in Hilbert Space

M.S. Livšic:
Commuting Nonselfadjoint Operators and
Collective Motions of Systems 1

L. Waksman:
Harmonic Analysis of Multi-Parameter
Semigroups of Contractions 40

COMMUTING NONSELFADJOINT OPERATORS AND COLLECTIVE MOTIONS OF SYSTEMS

Moshe S. Livšic

TABLE OF CONTENTS

Introduction

§1.	Single-Operator Colligations: Review of Basic Results	5
§2.	Commutative Colligations and Collective Motions	9
§3.	Characteristic (Transfer) Functions	14
§4.	The Output Realization of Colligations	17
§5.	Composit Colligations	22
§6.	Couplings and Resolutions of Commutative Colligations	23
§7.	A One-Dimensional Wave Equation	27
§8.	Symmetries	32

COMMUTING NONSELFADJOINT OPERATORS
AND COLLECTIVE MOTIONS OF SYSTEMS

M.S. Livšic
Department of Mathematics
Ben Gurion University of the Negev
Beer Sheva, Israel

Introduction

The spectral analysis of nonselfadjoint operators in the single operator case was developed during the fifties and sixties thanks to the efforts of many mathematicians. Then it was realized that this analysis forms a mathematical basis for the theory of open systems, interacting with the environment. In the light of the success of this theory, attempts have been made to create an analogous theory for several commuting operators, based on a generalization of the Nagy-Foias functional model to the case of several *independent* variables. It turned out, however, that this approach was ineffective. Moreover, the following generalization of the classical Cayley-Hamilton Theorem was obtained:

Two commuting operators with finite dimensional imaginary parts are connected, in generic case, by a certain algebraic equation. The degree of this equation does not exceed the dimension of the sum of ranges of imaginary parts.

This result shows that the genuine theory of commuting nonselfadjoint operators with finite dimensional imaginary parts is deeply connected with analytic functions on algebraic manifolds and it cannot be based only on functions of independent variables. Recent investigations of commuting operators was carried out in two directions. One of them deals with semigroups of shifts and it is presented in L.L. Waksman's monograph [13]. The other direction, which also is presented in this issue of "Lecture Notes" is based on operator colligations and collective motions of open systems [2-9]. In the case of collective motions all interacting systems distributed along the x-axis, form, at each given moment of time, one big spatial system. The corresponding input-output fields are compatible with internal collective states iff they satisfy certain partial differential equations of hyperbolic type. Actually, every given wave equation can be expressed as an external display of corresponding collective motions. In this way one can obtain, for instance, the Schrödinger, Klein-Gordon and Dirac equations. It turns out that the theory of commuting nonselfadjoint operators is deeply connected with the problem of wave dispersion in a medium: the above mentioned algebraic equation between commuting operators is nothing else but the corresponding dispersion law for the input-output waves. It is possible that the so called "quasi-particles" [1] can be described, at least from the mathematical point of view, as manifestations of collective motions. If the input field vanishes then the output manifestations of internal states behave like actions at a distance, decaying with respect to distance and depending also on time. If there exists a nonvanishing input wave, propagating along the x-axis, then quasi-particles are represented by a combined filed, consisting of the "ruling" input wave and of the all output manifestations, provoked by this wave.

§1. Single-Operator Colligations: Review of Basic Results

Let A be a bounded linear operator in a Hilbert space or in a finite dimensional space H. The closure G of the image of the difference $A-A^*$ is called the *nonhermitian* subspace of operator A. In many cases of interest this subspace is finite dimensional. The restriction of the type $\dim[\text{range}(A-A^*)] < \infty$ plays an important role. As an example let us take the Newton-Leibnitz integral

$$(Af)(x) = i \int_0^x f(s)ds, \quad (0 \leq x \leq \ell)$$

It is a nonselfadjoint operator in L_2 and the image of $2 \, \text{Im}(A)$:

$$\frac{1}{i}(A-A^*)f = \int_0^\ell f(s)ds, \quad (0 \leq x \leq \ell)$$

is the one-dimensional subspace of all constant functions in the interval $0 \leq x \leq \ell$. Moreover [2,3] the Newton-Leibnitz integral can be characterized up to the unitary equivalence by the properties:

1) $\dim[\text{range}(A-A^*)] = 1$
2) $\text{spectrum}(A) = \{0\}$

This example shows that the number $n = \dim G$ can be helpful for the classification of nonselfadjoint operators.

Many problems of Mathematical Physics lead to equation of the form

$$i \frac{\partial f}{\partial t} + Af = 0, \tag{1}$$

where $f(t) \in H$ is a state of the corresponding system. In physical applications the energy (or the number of particles) in the state f is proportional to the scalar product (f,f). If $A=A^*$ then the scalar product $(f(t),f(t))$ is a constant and in this case the system is said to be closed. In general the system interacts with environment and the corresponding operator A is not selfadjoint. It is always possible to represent the difference $A-A^*$ in the form

$$\frac{1}{i}(A-A^*) = \Phi^*\sigma\Phi \tag{2}$$

where $\Phi: H \to E$ is a linear mapping of H into a new Hilbert (or finite dimensional) space E, σ is a bounded *selfadjoint* operator in E and $\Phi^*: E \to H$ is the adjoint of Φ:

$$(\Phi f, u) = (f, \Phi^* u), \quad (f \in H, u \in E)$$

One of the representations of the form (2) can be derived in the following way:

$$\frac{1}{i}(A-A^*) = P_G \sigma P_G ,$$

where G is the nonhermition subspace, P_G is the orthogonal projection onto G,

$$\sigma = \frac{1}{i}(A-A^*)|_G$$

and $E=G$, $\Phi=P_G$.

The set $X = (A,H,\Phi,E,\sigma)$ which satisfies the condition (2) is called a *colligation*. In the following we assume that subspace E is finite dimensional: $\dim E = n < \infty$. The space H is called the *internal* space of the colligation and E — the *coupling* space.

With every colligation we associate an open system which is defined by equations of the form

$$i\frac{df}{dt} + Af = \Phi^*\sigma[u(t)] \tag{3}$$

$$v(t) = u(t) - i\Phi[f(t)] \tag{4}$$

where $u(t)$, $v(t) \in E$ are an input, and an output respectively. These relations have not been chosen arbitrarily. It is easy to check that for such a system the following law of metric (energy, number of particles) balance holds:

$$\frac{d}{dt}(f,f) = (\sigma u, u) - (\sigma v, v) , \tag{5}$$

where $(\sigma u, u)$ and $(\sigma v, v)$ can be interpreted as metric flows through the input and the output respectively.

Proof of the metric balance formula:

Using (3) and (4) we obtain

$$\frac{d}{dt}(f,f) = (\frac{df}{dt}, f) + (f, \frac{df}{dt}) = (iAf-i\phi^*\sigma(u), f) + (f, iAf-i\phi^*\sigma(u)) =$$
$$= (i(A-A^*)f, f) - i(\sigma(u), \Phi f) + i(\Phi f, \sigma(u))$$

and

$$(\sigma u, u) - (\sigma v, v) = (\sigma u, u) - (\sigma(u-i\Phi(f)), u-i\Phi(f)) =$$
$$= i(\Phi(f), \sigma(u)) - i(\sigma(u), \Phi(f)) - (\Phi^*\sigma\Phi(u), u)$$

The colligation condition (2) implies (5).

Let us consider a solution of equation (3) of the form
$$u(t) = u_0 e^{i\lambda t}, \quad f(t) = f_0 e^{i\lambda t}, \quad v(t) = v_0 e^{i\lambda t} \tag{6}$$

It is easy to see that
$$f_0 = (A-\lambda I)^{-1}\phi^* u_0 \tag{7}$$
$$v_0 = S(\lambda) u_0, \tag{8}$$

where
$$S(\lambda) = I - i\phi(A-\lambda I)^{-1}\phi^*\sigma(A) \tag{9}$$

and λ is a regular point of the resolvent.

Operator-function $S(\lambda)$ is said to be the *characteristic* (transfer) function of colligation (A, H, ϕ, E, σ). From (5) it follows for solution (6) that

$$i(\lambda-\lambda^*)(f_0, f_0) = (\sigma u_0, u_0) - (\sigma v_0, v_0) \tag{10}$$

This formula implies the following

Theorem 1

The characteristic function $S(\lambda)$ *has the following properties:*

$$(Im\lambda)[S^*(\lambda)\sigma S(\lambda) - \sigma] > 0, \quad (Im\lambda \neq 0) \tag{11}$$
$$S^*(\lambda)\sigma S(\lambda) = \sigma, \quad (Im\lambda = 0) \tag{12}$$

It can be proved the following important [2]

Factorization Theorem

Let $H = H_0 \supset H_1 \supset H_2 \supset \ldots H_{m-1} \supset H_m = 0$ *be a chain of invariant subspaces of operator* A *and* $H_k^\perp = H_k \ominus H_{k-1}$. *Then the characteristic function* $S(\lambda)$ *is a product*

$$S(\lambda) = S_m(\lambda) S_{m-1}(\lambda) \ldots S_2(\lambda) S_1(\lambda) \tag{13}$$

where $S_k(\lambda)$ $(k=1,2,\ldots,m)$ is the characteristic function of the colligation

$$X_k = (A|_{H_k^\perp}, H_k^\perp, P_k\Phi, E, \sigma) \tag{14}$$

(P_k is the orthogonal projection on $H_k^\perp$)

The colligation $X_k = P_k(X)$ is said to be the *projection* of X onto $H_k^\perp$. The colligation X is said to be the coupling of the projections X_k: $X = X_m \vee X_{m-1} \ldots \vee X_1$. To the coupling of colligations there correspondes the *chain coupling* [2,9] of corresponding systems: the output of each link coincides with the input of the next one.

It is easy to check that the subspace

$$\hat{H} = \text{span}\{A^k \phi^*(E)\} = \text{span}\{A^{*k} \phi^*(E)\}, \quad (0 \le k < \infty)$$

reduces A and that the restriction of A to the orthogonal complement $H_0 = H \ominus \hat{H}$ is selfadjoint. The subspace $\hat{H}$ is said to be the *principal* subspace of the colligation.

Theorem 2 [2]. *Let the space* E *and the selfadjoint operator* σ *be given. Assume that* $\det(\sigma) \ne 0$. *Then the characteristic function* $S(\lambda)$ *determines a corresponding colligation up to an unitary transformation of its principal component. The class* $\Omega(\sigma)$: An operator-function $W(\lambda): E \to E$ is said to be a function of the class $\Omega(\sigma)$ if it has the following properties:

1) $W(\lambda)$ is a meromorphic function in the open upper half plane $\text{Im}\lambda > 0$
2) $W(\lambda)$ is holomorphic in a neighbourhood $|\lambda| > a$ of $\lambda = \infty$ and $W(\infty) = I$.
3) $W^*(\lambda)\sigma W(\lambda) > \sigma$ $(\text{Im}\lambda > 0)$ (15)
4) $W^*(\lambda)\sigma W(\lambda) = \sigma$ $(\text{Im}\lambda = 0)$ (16)

The following theorem holds [2]:

Theorem 3

Let $\sigma = \sigma^*$ *be a given invertible operator in* E. *A given function* $W(\lambda)$ *belongs to the class* $\Omega(\sigma)$ *iff* $W(\lambda)$ *is the characteristic function of some colligation* $X = (A, H, \phi, E, \sigma)$.

If, for instance, $\dim E = 1$, then $S(\lambda)$ is a scalar function and we can assume that $\sigma = 1$. In this case the characteristic function $S(\lambda)$ can be represented in the following form [2,9]:

$$S(\lambda) = \prod_{k=1}^{N} \frac{\lambda - \lambda_k^*}{\lambda - \lambda_k} \cdot \exp\left[i \int_0^\ell \frac{ds}{\lambda - \alpha(s)}\right], \quad 0 \le N \le \infty \quad (17)$$

$(\text{Im}\lambda_k > 0, \Sigma \text{Im}\lambda_k < \infty)$, where $\alpha(s)$ is a real nondecreasing function. The corresponding operator $Af = f'$ can be representing (up to a unitary transformation) in the space $\ell_2 \oplus L_2$ in the following triangular form

$$f'_k = \lambda_k f_k + i \sum_{j=k+1}^{N} \beta_k \beta_j f_j + i\beta_k \int_0^{\ell} f(s)ds, \quad (j=1,2,\ldots,N) \tag{18}$$

$$f'(x) = \alpha(x)f(x) + i \int_0^x f(s)ds, \quad (0 \leq x \leq \ell)$$

where $\beta_k = (2 \text{ Im } \lambda_k)^{1/2}$. If $\dim H < \infty$, then $\ell=0$ and $N = \dim H$. If spectrum$(A) = \{0\}$ then $N=0$ and $\alpha(x) \equiv 0$. In this case the triangular model turns into the Newton-Leibnitz operator

$$(Af)(x) = i \int_0^x f(s)ds, \quad 0 \leq x \leq \ell. \tag{19}$$

The characteristic function in this case is

$$S(\lambda) = \exp(i \frac{\ell}{\lambda}) \tag{20}$$

For the generalization of formulas (17), (18) to the case $n > 1$ see [2].

§2. <u>Commutative Colligations and Collective Motions</u>

Let (A,B) be a pair of linear bounded operators in H. Define the subspaces

$$G_A = \overline{(A-A^*)H}, \quad G_B = \overline{(B-B^*)H}, \quad G = G_A + G_B \tag{21}$$

The subspace G is called the nonhermitian subspace of the pair (A,B). We assume in the following that $\dim G = n < \infty$. Let us consider, for example, the operators

$$(Af)(x) = i \int_0^x f(s)ds$$
$$\qquad\qquad\qquad (f \in L_2, 0 \leq x \leq \ell)$$
$$(Bf)(x) = \int_0^x (s-x)f(s)ds$$

It is easy to see that $B = A^2$ and

$$\frac{1}{i}(A-A^*)f = \int_0^{\ell} f(s)ds = (f,1)1$$

$$\frac{1}{i}(B-B^*)f = i \int_0^{\ell} (x-s)f(s)ds = i(f,1)x - i(f,x)1$$

Hence $G = \{c_1 + c_2 x\}$ is two dimensional

<u>Definition</u>. Let H, E be Hilbert spaces, (A,B)-bounded linear operators in H; $\sigma(A), \sigma(B)$-bounded selfadjoint operators in E, Φ - a bounded linear mapping of H into E. A set

$$X = (A,B;H,\Phi,E;\sigma(A),\sigma(B)) \tag{22}$$

is said to be a colligation if

$$\frac{1}{i}(A-A^*) = \Phi^*\sigma(A)\Phi, \quad \frac{1}{i}(B-B^*) = \Phi^*\sigma(B)\Phi \qquad (23)$$

If range $\Phi = E$ and Ker $\sigma(A) \cap$ Ker $\sigma(B) = 0$ the colligation is called a *strict* colligation. A colligation is said to be *commutative* if $AB=BA$. An arbitrary given pair (A,B) can be embedded in a strict colligation with $E=G$, $\Phi=P_G$, $\sigma(A) = \frac{1}{i}(A-A^*)\big|_G$, $\sigma(B) = \frac{1}{i}(B-B^*)\big|_G$.

If $AB=BA$ then the ranges of selfadjoint operators

$$\frac{1}{i}(AB^*-BA^*) = i[(B-B^*)A^* - (A-A^*)B^*] \qquad (24)$$

$$\frac{1}{i}(B^*A-A^*B) = i[(B-B^*)A - (A-A^*)B] \qquad (25)$$

belong to the nonhermitian subspace

$$G = \text{range}(A-A^*) + \text{range}(B-B^*) \qquad (26)$$

Remark. If X is strict then $\Phi^*E=G$, $\Phi G=E$ and dim E = dim G. Indeed, relation (23) imply that range$(A-A^*)$ = range$\Phi^*\sigma(A)$, range$(B-B^*)$ = range$\Phi^*\sigma(B)$ and $G = \Phi^*(\text{range}\sigma(A)+\text{range}\sigma(B)) = \Phi^*E$. If one assumes that $(\Phi G, U_0)=0$ then $(G,\Phi^*U_0)=0$ and $\Phi^*U_0=0$. Hence $(\Phi H, U_0) = (H,\Phi^*U_0) = 0$ and $U_0 = 0$.

Corollary. If X is strict and commutative then there are two selfadjoint operators γ and $\widetilde{\gamma}$ in E, satisfying the following conditions:

$$\frac{1}{i}(AB^*-BA^*) = \Phi^*\gamma\Phi \qquad (27)$$

$$\frac{1}{i}(B^*A-A^*B) = \Phi^*\widetilde{\gamma}\Phi , \qquad (28)$$

where $\Phi_0 = \Phi|G$.

Remark. The equalities (24) and (23) imply

$$2 \text{ Im}(AB^*) = \Phi^*[\sigma(A)\Phi B^* - \sigma(B)\Phi A^*] ,$$

$$2 \text{ Im}(B^*A) = \Phi^*[\sigma(A)\Phi B - \sigma(B)\Phi A]$$

Hence for strict colligations we obtain:

$$\sigma(A)\Phi B^* - \sigma(B)\Phi A^* = \gamma\Phi$$

$$\sigma(A)\Phi B - \sigma(B)\Phi A = \widetilde{\gamma}\Phi ,$$

Then substituting and using (23) we obtain

$$i[\sigma(A)\Phi\Phi^*\sigma(B) - \sigma(B)\Phi\Phi^*\sigma(A)]\Phi = (\widetilde{\gamma}-\gamma)\Phi$$

Hence, for strict colligation the relation

$$\widetilde{\gamma} - \gamma = i[\sigma(A)\Phi\Phi^*\sigma(B) - \sigma(B)\Phi\Phi^*\sigma(A)]$$

is valid.

Operators γ, $\tilde{\gamma}$ play an important role but the condition $\Phi(H) = E$ is too restrictive: the projection of a strict colligation on a subspace is not always strict. Moreover, in the simplest case when dim $H = 1$, colligations are not strict unless dim $E = 1$. To overcome these difficulties we use the notion of a regular colligation, introduced by N. Kravitsky [5].

<u>Definition</u>. A commutative colligation is said to be *regular* if there exists a selfadjoint operator γ in E such that the equality

$$\sigma(A)\Phi B^* - \sigma(B)\Phi A^* = \gamma\Phi \tag{29}$$

is valid. Let us define $\tilde{\gamma}$ as follows:

$$\tilde{\gamma} = \gamma + i[\sigma(A)\Phi\Phi^*\sigma(B) - \sigma(B)\Phi\Phi^*\sigma(A)] \tag{29'}$$

It is easy to check that $\tilde{\gamma}$ satisfies the following condition

$$\sigma(A)\Phi B - \sigma(B)\Phi A = \tilde{\gamma}\Phi \tag{30}$$

Analogously from (30), (29') it follows (29). It is evident that strict colligations are regular. The operators γ and $\tilde{\gamma}$ are defined uniquely for strict colligations but for regular colligations it is not so. We will include operators γ and $\tilde{\gamma}$ in the notation of a regular colligation

$$X = (A,B;H,\Phi,E;\sigma(A),\sigma(B),\gamma,\tilde{\gamma}) ,$$

where γ and $\tilde{\gamma}$ are connected by the so called "linkage" equality (29'). They satisfy the regularity conditions (29), (30). The regularity conditions can be written in a determinantal form

$$\begin{vmatrix} \sigma(A)\Phi, & A^* \\ \sigma(B)\Phi, & B^* \end{vmatrix} = \gamma\Phi , \quad \begin{vmatrix} \sigma(A)\Phi, & A \\ \sigma(B)\Phi, & B \end{vmatrix} = \tilde{\gamma}\Phi \tag{31}$$

and the linkage equation has the form

$$\begin{vmatrix} \sigma(A)\Phi, & \Phi^*\sigma(A) \\ \sigma(B)\Phi, & \Phi^*\sigma(B) \end{vmatrix} = \frac{1}{i}(\tilde{\gamma}-\gamma) \tag{32}$$

Let us assume now that an input, a state and an output depend also on a spatial coordinate x ($x_0 \leq x \leq x_1$).

<u>Definition</u>. An input $u(t,x)$, a state $f(t,x)$ and an output $v(t,x)$ are said to be a *collective* input, state and output respectively if they satisfy equations of the form:

$$i\frac{\partial f}{\partial t} + Af = \Phi^*\sigma(A)[u(t,x)] , \tag{33}$$

$$i\frac{\partial f}{\partial x} + Bf = \Phi^*\sigma(B)[u(t,x)] \tag{34}$$

$$v(t,x) = u(t,x) - i\Phi[f(t,x)], \qquad (35)$$

where

$$X = (A,B;H,\Phi,E;\sigma(A),\sigma(B)) \qquad (36)$$

is a colligation, i.e.

$$\frac{1}{i}(A-A^*) = \Phi^*\sigma(A)\Phi$$
$$\frac{1}{i}(B-B^*) = \Phi^*\sigma(B)\Phi. \qquad (37)$$

So, at the arbitrary fixed point x $(x_0 \leqslant x \leqslant x_1)$ the equation (33) defines a temporal system. The input and the internal state depend also on x. In the case of collective motions all temporal systems, distributed along the x-axis, at the same moment of time behave like one big spatial system: the equation (34) defines a connection between the internal states at different points of the x-axis. For collective motions the following laws of metric balance hold:

$$\frac{\partial}{\partial t}(f,f) = (\sigma(A)u,u) - (\sigma(A)v,v) \qquad (38)$$

$$\frac{\partial}{\partial x}(f,f) = (\sigma(B)u,u) - (\sigma(B)v,v) \qquad (39)$$

Assuming that equations (33) and (34) are consistent for arbitrary initial conditions $f(t_0,x_0)=f_0$ we conclude that $AB=BA$. On the line

$$t = \xi_1\tau+t_0, \qquad x = \xi_2 c\tau+x_0 \qquad (40)$$

where c denotes the light speed in the vacuum, equations (33), (34) imply

$$i\frac{df}{d\tau} + (\vec{\xi}\cdot\vec{A})f = \Phi^*\sigma(\vec{\xi}\cdot\vec{A})(u), \qquad (41)$$

$$v = u - i\Phi(f), \qquad (42)$$

where $\vec{\xi}\cdot\vec{A} = \xi_1 A_1 + \xi_2 A_2$, $A_1 = A$, $A_2 = cB$.

The system (41) in the direction $\vec{\xi} = (\xi_1,\xi_2)$ corresponds to the motion $x = V\cdot(t-t_0)+x_0$ of the point x along the x-axis with the constant velocity $V = \frac{\xi_2}{\xi_1}c$. If $\xi_1=1$, $\xi_2=0$ then $V=0$ and $x=x_0$. In this case we obtain the temporal system at fixed point $x=x_0$. If $\xi_1=0$, $\xi_2=1$ then $V=\infty$ and $t=t_0$, which corresponds to the spatial system at fixed moment of time $t=t_0$. Thus, in the case of collective motions there exists a family of open systems related to motions $x = Vt+x_0$ along the x-axis. In particular, the spatial system corresponds to the *infinite* speed of the point x, which is an *immediate action at a distance: the spatial system defines a solid frame for all possible motions of temporal systems, distributed along the* x-*axis*.

Theorem 4

Assume that X *is a strict commutative colligation. Then an input* u(t,x) *is a collective input if and only if* u(t,x) *satisfies the* PDE

$$\sigma(B)\frac{\partial u}{\partial t} = \sigma(A)\frac{\partial u}{\partial x} - i\gamma u \tag{43}$$

The corresponding output satisfies the equation

$$\sigma(B)\frac{\partial v}{\partial t} = \sigma(A)\frac{\partial v}{\partial x} - i\tilde{\gamma}v \tag{44}$$

Proof. We use integrability conditions

$$\frac{\partial^2 f}{\partial t \partial x} = \frac{\partial^2 f}{\partial x \partial t}$$

Equations (33), (34) imply

$$\frac{\partial}{\partial x}(Af-\Phi^*\sigma(A)u) = \frac{\partial}{\partial t}(Bf-\Phi^*\sigma(B)u)$$

Hence

$$A\frac{\partial f}{\partial x} - B\frac{\partial f}{\partial t} = \Phi^*[\sigma(A)\frac{\partial u}{\partial x} - \sigma(B)\frac{\partial u}{\partial t}] \tag{45}$$

Equations (33), (34) also imply

$$A\frac{\partial f}{\partial x} - B\frac{\partial f}{\partial t} = i[B\Phi^*\sigma(A) - A\Phi^*\sigma(B)]u$$

Using (29) we obtain

$$A\frac{\partial f}{\partial x} - B\frac{\partial f}{\partial t} = i\Phi^*\gamma \tag{46}$$

Equalities (45), (46) imply

$$\Phi^*[\sigma(A)\frac{\partial u}{\partial x} - \sigma(B)\frac{\partial u}{\partial t} - i\gamma u] = 0$$

which implies for strict colligations the equation (43). For v=u-iΦf we obtain

$$\sigma(A)\frac{\partial v}{\partial x} - \sigma(B)\frac{\partial v}{\partial t} - i\tilde{\gamma}v =$$

$$= \sigma(A)\frac{\partial u}{\partial x} - \sigma(B)\frac{\partial u}{\partial t} - i\sigma(A)\Phi\frac{\partial f}{\partial x} + i\sigma(B)\Phi\frac{\partial f}{\partial t} - i\tilde{\gamma}u - \tilde{\gamma}\Phi f =$$

$$= i(\gamma-\tilde{\gamma})u + [\sigma(A)\Phi B - \sigma(B)\Phi A]f - [\sigma(A)\Phi\Phi^*\sigma(B) - \sigma(B)\Phi\Phi^*\sigma(A)]u - \tilde{\gamma}\Phi f .$$

Using (30) in the form

$$\sigma(A)\Phi B - \sigma(B)\Phi A = \tilde{\gamma}\Phi \tag{47}$$

and

$$\tilde{\gamma}-\gamma = i(\sigma(A)\Phi\Phi^*\sigma(B) - \sigma(B)\Phi\Phi^*\sigma(A)) \tag{48}$$

we obtain

$$\sigma(A)\frac{\partial v}{\partial x} - \sigma(B)\frac{\partial v}{\partial t} - i\tilde{\gamma}v = i(\gamma-\tilde{\gamma})u + \tilde{\gamma}\Phi u + i(\tilde{\gamma}-\gamma)u - \tilde{\gamma}\Phi f = 0$$

Remarks: 1) Theorem 4 implies that γ corresponds to the input and $\tilde{\gamma}$ corresponds to the output in a natural way. We will denote $\gamma = \gamma^{in}$ and $\tilde{\gamma} = \gamma^{out}$. 2) If $\sigma(B) > 0$ then PDE (43), (44) are of *hyperbolic* type. Let us consider collective motions of the form

$$u = u_0 e^{i(\lambda t + \mu x)}, \quad f = f_0 e^{i(\lambda t + \mu x)}, \quad v = v_0 e^{i(\lambda t + \mu x)} \qquad (49)$$

Theorem 4 implies that u and v are solutions of equations (43) and (44) respectively. Hence u_0 and v_0 are solutions of algebraic equations

$$(\lambda\sigma(B) - \mu\sigma(A) + \gamma^{in})u_0 = 0 \qquad (50)$$

$$(\lambda\sigma(B) - \mu\sigma(A) + \gamma^{out})v_0 = 0 \qquad (51)$$

Later we will prove the following important equality:

$$\det(\lambda\sigma(B) - \mu\sigma(A) + \gamma^{in}) \equiv \det(\lambda\sigma(B) - \mu\sigma(A) + \gamma^{out}) \qquad (52)$$

The polynomial

$$D(\lambda,\mu) = \det(\lambda\sigma(B) - \mu\sigma(A) + \gamma^{in}) \qquad (53)$$

is said to be the *discriminant* of the colligation. The corresponding algebraic curve

$$\Gamma = \{(\lambda,\mu) \in C_2, D(\lambda,\mu) = 0\}$$

is said to be the *discriminant curve*.
The subspaces

$$E^{in}(M) = \operatorname{Ker}(\lambda\sigma(B) - \mu\sigma(A) + \gamma^{in}), \qquad (54)$$

$$E^{out}(M) = \operatorname{Ker}(\lambda\sigma(B) - \mu\sigma(A) + \gamma^{out}) \qquad (55)$$

where $M = (\lambda,\mu)$ is an arbitrary point of the curve Γ are said to be the joint input and the joint output subspaces respectively. Equations (50), (51) imply that $u_0 \in E^{in}(M)$, $v_0 \in E^{out}(M)$. In the case $\sigma(B) > 0$ to each real value μ there correspond n real roots $\lambda_1(\mu), \ldots, \lambda_n(\mu)$ of the equation

$$D(\lambda,\mu) = 0 \qquad (56)$$

Hence, to each real value μ there correspond n plane collective waves of the form (49).

§3. Characteristic (Transfer) Functions

Let us consider special motions

$$u = u_0 e^{iz\tau}, \quad f = f_0 e^{iz\tau}, \quad v = v_0 e^{iz\tau} \qquad (57)$$

of the system (41) on the line $t=\xi_1\tau$, $x=\xi_2 c\tau$. It is evident that

$$f_0 = (\alpha A+\beta B-zI)^{-1}\Phi^*(\alpha\sigma(A)+\beta\sigma(B)) \tag{58}$$

$$v_0 = S(\alpha,\beta,z)u_0 ,$$

where $\alpha=\xi_1$, $\beta=c\xi_2$ and

$$S(\alpha,\beta,z) = I-i\Phi(\alpha A+\beta B-zI)^{-1}\Phi^*(\alpha\sigma(A)+\beta\sigma(B)) \tag{59}$$

The operator function $S(\alpha,\beta,z)$ of variables α,β,z is said to be the *complete characteristic* (transfer) *function* (CCF) of the colligation X: it is a regular analytic function of z and of parameters α, β at all points for which there exist the resolvent. It is evident that $S(\rho\alpha,\rho\beta,\rho z) = S(\alpha,\beta,z)$. If, in particular, $\alpha=1$, $\beta=0$, $z=\lambda$ then

$$S(1,0,\lambda) = I-i\Phi(A-\lambda I)^{-1}\Phi^*\sigma(A) \tag{60}$$

is the temporal CF. In the case $\alpha=0$, $\beta=1$, $z=\mu$ we obtain the spatial CF

$$S(0,1,\mu) = I-i\Phi(B-\mu I)^{-1}\Phi^*\sigma(B) \tag{61}$$

Theorem 5. *If* $z = \alpha\lambda+\beta\mu$ *and* $D(\lambda,\mu) = 0$ *then the restriction of the CCF to the joint input subspace* $E^{in}(M)$, $M(\lambda,\mu) \in \Gamma$:

$$S(\alpha,\beta,\alpha\lambda+\beta\mu)\Big|_{E^{in}(M)}$$

depends only on the point $M(\lambda,\mu)$ *of the discriminant curve* Γ

Proof. Relations (29), (50) imply for $u_0 \in E^{in}(M)$

$$\begin{aligned}(B\Phi^*\sigma(A)-A\Phi^*\sigma(B))u_0 &= \Phi^*\gamma^{in}u_0 \\ \Phi^*(\mu\sigma(A)-\lambda\sigma(B))u_0 &= \Phi^*\gamma^{in}u_0\end{aligned} \tag{62}$$

Hence

$$(B-\mu I)\Phi^*\sigma(A)u_0 = (A-\lambda I)\Phi^*\sigma(B)u_0$$

and

$$(A-\lambda I)^{-1}\Phi^*\sigma(A)u_0 = (B-\mu I)^{-1}\Phi^*\sigma(B)u_0 = f_0 \tag{63}$$

Then

$$\begin{aligned}(A-\lambda I)f_0 &= \Phi^*\sigma(A)u_0 \\ (B-\mu I)f_0 &= \Phi^*\sigma(B)u_0\end{aligned}$$

and

$$[\alpha(A-\lambda I)+\beta(B-\mu I)]f_0 = \Phi^*(\alpha\sigma(A)+\beta\sigma(B))u_0$$

Thus,

$$f_0 = [\alpha A + \beta B - (\alpha\lambda + \beta\mu)1]^{-1} \Phi^* (\alpha\sigma(A) + \beta\sigma(B)) u_0 \tag{64}$$

does not depend on α, β.

<u>Remark</u>. The last relation implies that the internal state f_0 which corresponds to input $u_0 \in E^{in}(M)$ depends only on the point $M(\lambda,\mu)$ on the discriminant curve Γ (and on u_0).

<u>Definition</u>. The operator

$$\hat{S}(M) = S(\alpha,\beta,\alpha\lambda+\beta\mu)\big|_{E^{in}(M)} \tag{65}$$

is said to be the *joint characteristic function* (JCF) of commutative colligation X.

As a function of $M(\lambda,\mu)$ it is defined on the discriminant curve Γ and as an operator it is defined on $E^{in}(M)$.

Now we will cite some basic results on characteristic function of commuting colligations. Fir proofs see [8].

<u>Lemma</u>. For regular commuting colligations the following relation holds:

$$(\lambda\sigma(B) - \mu\sigma(A) + \gamma^{out}) S(\alpha,\beta,\alpha\lambda+\beta\mu) =$$
$$= \tilde{S}(\alpha,\beta,\alpha\lambda+\beta\mu)(\lambda\sigma(B) - \mu\sigma(A) + \gamma^{in}) , \tag{66}$$

where $S(\alpha,\beta,z)$ is the CCF and

$$\tilde{S}(\alpha,\beta,z) = I - i(\alpha\sigma(A) + \beta\sigma(B))\Phi(\alpha A + \beta B - zI)^{-1}\Phi^* \tag{67}$$

<u>Corollary</u>. For regular colligations the following important *law of conservation* holds:

$$\det(\lambda\sigma(B) - \mu\sigma(A) + \gamma^{in}) \equiv \det(\lambda\sigma(B) - \mu\sigma(A) + \gamma^{out}) \tag{68}$$

<u>Corollary</u>. The joint characteristic function maps $E^{in}(M)$ into $E^{out}(M)$.

<u>Theorem 6</u>. The CCF of a regular colligation with $\sigma(B) > 0$ is determined uniquely by $\sigma(A), \sigma(B), \gamma^{in}$ and the JCF.

<u>Corollary</u>. A regular commutative colligation with $\sigma(B) > 0$ is determined up to a unitary transformation of its principal subspace by the JCF $\hat{S}(M)$: if

$$X' = (A',B';H',\Phi',E;\sigma(A),\sigma(B),\gamma^{in},\gamma^{out})$$
$$X'' = (A'',B'';H'',\Phi'',E;\sigma(A),\sigma(B),\gamma^{in},\gamma^{out})$$

are two regular colligations with the same JCF then there exists an isometric transformation T of the principal subspace $\tilde{H}'$ onto the principal subspace $\tilde{H}''$ such that $A'' = TA'T^{-1}$, $B'' = TB'T^{-1}$, $\Phi' = \Phi''T$.

<u>Corollary</u>. A CCF $S(\alpha,\beta,z)$ is determined uniquely by $\sigma(B) > 0$, $\sigma(A)$, γ^{in} and its values $S(\alpha_0,\beta_0,z)$ at an arbitrary fixed point $(\alpha_0,\beta_0) \neq 0$.

<u>Remark</u>.

Putting $\lambda=-\omega$, $\mu=k$ in (49):
$$u = u_0 e^{i(kx-\omega t)}, \quad f = f_0 e^{i(kx-\omega t)}, \quad v = v_0 e^{i(kx-\omega t)}$$
we obtain plain waves, where ω is the frequency and k is the wave number or spatial "frequency". Hence, there exists a simple relation between the discriminant curve $D(\lambda,\mu) = 0$ and the dispersion law $F(\omega,k) = 0$: $F(\omega,k) = D(-\omega,k)$. The value $z = -\alpha\omega+\beta k$ in the definition (65) of JCF corresponds to the motion $t=\alpha\tau$, $x=\beta\tau$. Hence
$$S(\alpha,\beta,\alpha\lambda+\beta\mu)\Big|_{E^{in}(M)} = S(1,v,-\omega+vk)\Big|_{E^{in}(M)}, \tag{69}$$
where $V=\beta/\alpha$ is the speed of point x and $\omega-vk = -z$ is the frequency related to the motion $x=vt$ along the x-axis. The meaning of Theorem 5 is: In the case of collective motions the transfer functions (and the internal states) on the joint input subspace are the same for all possible motions $x=vt$ along the x-axis. In particular, the spatial and the temporal CF coincide on $E^{in}(M)$:
$$S(1,0,-\omega)\Big|_{E^{in}(M)} = S(0,1,k)\Big|_{E^{in}(M)} \tag{70}$$

§4. The Output Realization of Colligations

Let $u=0$ in (33), (34). Then using variables τ and ξ instead of t and x we obtain

$$(i\frac{\partial}{\partial\tau} + A)f(\tau,\xi) = 0$$

$$(i\frac{\partial}{\partial\xi} + B)f(\tau,\xi) = 0 \tag{71}$$
$$f(0,0) = f_0$$
$$v(\tau,\xi) = -i\Phi(f)$$

The solution of equations has the following form
$$f(\tau,\xi) = e^{i(\tau A+\xi B)}f_0 \tag{72}$$

The relation (39) implies
$$\frac{\partial}{\partial\xi}(f,f) = -(\sigma(B)v,v) \tag{73}$$

We assume in the following that $\sigma(B) > 0$. Then (73) implies that

$(f(0,\xi), f(0,\xi))$ is a decreasing function of ξ $(0 \leq \xi < \infty)$. We assume also that operator B satisfies the condition [10]:

$$\lim_{\xi \to +\infty}(e^{i\xi B}f_0, e^{i\xi B}f_0) = 0 \tag{74}$$

for every f_0 $(f_0 \in H)$. Then the relation (73) implies

$$(f_0, f_0) = \int_0^\infty (\sigma(B)v[f_0](0,\xi), v[f_0](0,\xi))d\xi, \tag{75}$$

where

$$v[f_0](\tau,\xi) = \Phi e^{i(\tau A + \xi B)}f_0 \tag{76}$$

The regularity condition (30) can be written in the form

$$\sigma(A)\Phi B - \sigma(B)\Phi A = \gamma^{out}\Phi \tag{77}$$

From (76), (77) we conclude that $v[f_0]$ satisfies the PDE:

$$[\sigma(B)\frac{\partial}{\partial \tau} - \sigma(A)\frac{\partial}{\partial \xi} + i\gamma^{out}]v = 0 \tag{78}$$

The relation (76) defines a mapping of H onto a subset $\tilde{H}$ of solutions of PDE (78). All the functions $v[f_0](\tau,\xi)$, belonging to H, are entire analytic functions. We introduce in $\tilde{H}$ the following scalar product:

$$< v_1, v_2 > = \int_0^\infty (\sigma(B)v_1(0,\xi), v_2(0,\xi))d\xi \tag{79}$$

The relation (75) implies that $v[f_0](\tau,\xi)$ defines an isometric mapping of H onto $\tilde{H}$. The above considerations lead us to the following.

Theorem 7. *If* X *is a regular commutative colligation with* $\sigma(B) > 0$ *and*

$$\lim_{x \to +\infty} e^{ixB}f_0 = 0, \quad (f_0 \in H) \tag{80}$$

then X *is unitary equivalent to the colligation*

$$X = (-i\frac{\partial}{\partial \tau}, -i\frac{\partial}{\partial \xi}; \tilde{H}, \tilde{\Phi}, E; \sigma(A), \sigma(B)) \tag{81}$$

where $\tilde{H}$ *is a set of solutions* $v(\tau,\xi)$ *of the equation*

$$\sigma(B)\frac{\partial v}{\partial \tau} - \sigma(A)\frac{\partial v}{\partial \xi} - i\gamma^{out}v = 0 \tag{82}$$

such that
1) $\tilde{H}$ *is a Hilbert space with respect to the scalar product* (79)
2) *If* $v(\tau,\xi)$ *belongs to* $\tilde{H}$ *then* $\frac{\partial v}{\partial \tau}$ *and* $\frac{\partial v}{\partial \xi}$ *belong to* H
3) *Operators* $\frac{\partial}{\partial \tau}$, $\frac{\partial}{\partial \xi}$ *are bounded in* H

The mapping Φ *is defined by*

$$\Phi v(\tau,\xi) = v(0,0) \tag{83}$$

We have already proved part 1). The equality (76) implies

$$\tilde{A}v[f_0] = v[Af_0] = -i\frac{\partial v[f_0]}{\partial \tau}$$

$$\tilde{B}v[f_0] = v[Bf_0] = -i\frac{\partial v[f_0]}{\partial \xi} \tag{84}$$

$$\tilde{\Phi}v[f_0] = \Phi f_0 = v[f_0](0,0)$$

which proves the theorem.

<u>Corollary</u>. The operators $\tilde{A}$, $\tilde{B}$ are infinitesimal shifts. To the group of transformations

$$T(t,x) = e^{i(tA+xB)} \tag{85}$$

in H there corresponds the group of shifts

$$\tilde{T}(t,x)v(\tau,\xi) = v(t+\tau,x+\xi) \tag{86}$$

in $\tilde{H}$.

<u>Definition</u>. The functions $v[f_0](t,\xi)$ and $v[t_0](0,\xi)$ are said to be the *output representation* and the *mode* of an element f_0 correspondingly. If a mode $v_0(\xi)$ is given then the corresponding output representation $v(\tau,\xi)$ is defined uniquely by PDE (82) and the initial condition $v(0,\xi) = v_0(\xi)$. The transformation

$$v[f_0](0,\xi) = \Phi e^{i\xi B} f_0 \quad (f_0 \in H, 0 \le \xi < \infty)$$

is an isometric mapping of H onto the "mode-space" $\tilde{H}_0$ with respect to scalar product (79). Hence each element $f \in H$ is represented by its mode $v_f(\xi)$ $(0 \le \xi < \infty)$ which is an entire analytic function of ξ. On the subspace $\tilde{H}$ the operator $-i\frac{\partial}{\partial \xi} = \tilde{B}$ is bounded. From equation (82) it follows that operators $\tilde{A} = -i\frac{\partial}{\partial \tau}$, $\tilde{B} = -i\frac{\partial}{\partial \xi}$ on the subspace $\tilde{H}$ satisfy the equation

$$\sigma(B)(-i\frac{\partial}{\partial t}) - \sigma(A)(-i\frac{\partial}{\partial \xi}) + \gamma^{out}] = 0 \tag{87}$$

which implies that

$$D(-i\frac{\partial}{\partial \tau}, -i\frac{\partial}{\partial \xi}) = 0, \tag{88}$$

where

$$D(\lambda,\mu) = \det(\lambda\sigma(B) - \mu\sigma(A) + \gamma^{out}) \tag{89}$$

is the discriminatn polynomial.

Using the unitary equivalence between the pair (A,B) and the pair $(\tilde{A} = -i\frac{\partial}{\partial \tau}, \tilde{B} = -i\frac{\partial}{\partial \xi})$ we conclude that the pair (A,B) satisfies the following algebraic equation

$$D(A,B) = 0 \qquad (90)$$

Thus, we arrived to the following generalization of the Cayley-Hamilton Theorem:

Theorem 8. If X is commutative regular colligation such that

$$\lim_{x \to +\infty} e^{ixB} f = 0, \qquad (\sigma(B) > 0, f \in H) \qquad (91)$$

then the pair (A,B) satisfies the algebraic equation $D(A,B) = 0$.

Remarks. 1) The relation (90) can be proved without conditions (91) in the following form [5,6]:

$$D(A,B)\big|_{\hat{H}} = 0, \qquad (92)$$

where $\hat{H}$ is the principal subspace of X:

$$\hat{H} = \text{span}\{A^j B^k \phi^*(E)\}, \qquad (j,k = 0,1,2,\ldots) \qquad (93)$$

On the subspace $H \ominus \hat{H}$ operators A, B are selfadjoint.

2) It is impossible to create an impulse within the mode-subspace $\tilde{H}_0$. It means that it is impossible to construct in $\tilde{H}_0$ a sequence $v_n(x)$ $(0 \leq x < \infty)$, such that

$$[v_n, v_n] = \int_0^\infty (\sigma(B) v_n(x), v_n(x)) dx = 1 \qquad (94)$$

and

$$\lim_{n \to \infty} \int_{x_0 - \varepsilon_n}^{x_0 + \varepsilon_n} (\sigma(B) v_n(x), v_n(x)) dx = 1 \qquad (95)$$

where $\varepsilon_n \to 0$, $(\varepsilon_n > 0)$ and x_0 is an aribtrary given point in the region $0 \leq x_0 < \infty$. Indeed, if $v_n \in H_0$ and $[v_n, v_n] = 1$ then

$$v_n(x) = \Phi e^{ixB} f_n$$

and $[v_n, v_n] = (f_n, f_n) = 1$. The condition $\sigma(B) > 0$ implies

$$\frac{d}{dx} (e^{ixB} f, e^{ixB} f) = -(\sigma(B) v, v) \leq 0$$

Hence, $||e^{ixB}|| \leq 1$ if $x \geq 0$. Then

$$(\sigma(B) v_n(x), v_n(x)) = (\sigma(B) \Phi e^{ixB} f_n, \Phi e^{iBx} f_n) \leq ||\sigma(B)|| \cdot ||\Phi||$$

and

$$\int_{x_0 - \varepsilon_n}^{x_0 + \varepsilon_n} (\sigma(B) v_n(x), v_n(x)) dx \leq 2\varepsilon_n ||\sigma(B)|| \cdot ||\Phi||$$

Hence

$$\lim_{n\to\infty} \int_{x_0-\varepsilon_n}^{x_0+\varepsilon_n} (\sigma(B)v_n(x),v_n(x))dx = 0 \qquad (96)$$

Conclusions

Collective motions in the case $\sigma(B) > 0$ have the following properties:

1) An input $u(t,x)$ satisfies the PDE

$$(\sigma(B)\frac{\partial}{\partial t} - \sigma(A)\frac{\partial}{\partial x} + i\gamma^{in})u = 0 \qquad (97)$$

of a hyperbolic type. The distribution of values $u(t_0,x)$ ($-\infty < x < \infty$) at the initial moment of time may be an "arbitrary" function of x. An initial impulse on the input can be created and it propagates with a finite velocity (close-range action)

2) An internal state f can be represented by its outputs at all points of the region $0 \leqslant x < \infty$:

$$(f,f) = \int_0^\infty (\sigma(B)v(x),v(x))dx , \qquad (98)$$

where

$$v(x) = \Phi e^{ixB} f . \qquad (99)$$

The corresponding output representation $v(t,x)$ is a solution of PDE (44). The subspace $\tilde{H}_0$ of all modes (all initial conditions) is restricted by the requirement of continuity of the operator $\frac{\partial}{\partial x}$. It is impossible to create an initial impulse within the space $\tilde{H}_0$. Thus, *collective states are actions at a distance*. Inputs, compatible with internal collective motions, must be waves, satisfying wave equation (97). In the limit case $\Phi \to 0$ all modes (99) vanish but the wave equation (97) is not changing.

Remark. If quasi-particles which play an important role in modern physics are nothing else but collective states of spatio-temporal systems then they may appear in the space as output displays of internal states. They are actions at a distance, decaying with respect to the distance and also depending on the time. The field at the input vanishes in this case. However, there can exist also an input, wave, propagating along the x-axis. It is compatible with collective states iff it satisfies the PDE at the input. Particles in this case are represented by a combined field consisting of the "ruling" wave at the input and of the output displays of internal states, provoked by the ruling wave.

§5. Composit Colligations

Let us consider, for example, the following pair of 2×2 matrices.

$$A = \begin{pmatrix} c\mu_0 & -a \\ -a & -c\mu_0 \end{pmatrix}, \quad B = \mu_0 \begin{pmatrix} 1 & 0 \\ 0 & 1 \end{pmatrix} \tag{100}$$

where $c > 0$, $a \geq 0$ and $\mu_0 = \alpha_0 + \frac{i}{2}\beta_0^2$ is a nonreal number ($-\infty < \alpha < \infty$, $\beta_0 > 0$). Then

$$2\text{Im}(A) = \frac{1}{i}(A-A^*) = c\beta_0^2 \begin{pmatrix} 1 & 0 \\ 0 & -1 \end{pmatrix}, \quad 2\text{Im}(B) = \beta_0^2 \begin{pmatrix} 1 & 0 \\ 0 & 1 \end{pmatrix} \tag{101}$$

$$2\text{Im}(AB^*) = \frac{1}{i}(A\mu_0^* - \mu_0 A^*) =$$

$$= \frac{1}{i}\left[\left(c\mu_0\begin{pmatrix} 1 & 0 \\ 0 & -1 \end{pmatrix} - a\begin{pmatrix} 0 & 1 \\ 1 & 0 \end{pmatrix}\right)\mu_0^* - \left(c\mu_0^*\begin{pmatrix} 1 & 0 \\ 0 & -1 \end{pmatrix} - a\begin{pmatrix} 0 & 1 \\ 1 & 0 \end{pmatrix}\right)\mu_0\right] =$$

$$= a\beta_0^2 \begin{pmatrix} 0 & 1 \\ 1 & 0 \end{pmatrix} \tag{102}$$

It is evident that

$$\text{Im}(B^*A) = \text{Im}(AB^*) \tag{103}$$

Hence, the pair (A,B) can be embedded in the strict colligation X with

$$\sigma(A) = a\begin{pmatrix} 1 & 0 \\ 0 & -1 \end{pmatrix}, \quad \sigma(B) = \frac{a}{c}\begin{pmatrix} 1 & 0 \\ 0 & 1 \end{pmatrix}, \quad \gamma^{in} = \gamma^{out} = \frac{a^2}{c}\begin{pmatrix} 0 & 1 \\ 1 & 0 \end{pmatrix} \tag{104}$$

and $H = \mathbb{C}_2$, $E = \mathbb{C}_2$, $\Phi f = \beta_0(\frac{c}{a})^{\frac{1}{2}}f$, $\Phi^* u = \beta_0(\frac{c}{a})^{\frac{1}{2}}u$.

The corresponding PDE on the input and the output coincide:

$$\begin{pmatrix} 1 & 0 \\ 0 & 1 \end{pmatrix}\frac{\partial u}{\partial t} - c\begin{pmatrix} 1 & 0 \\ 0 & -1 \end{pmatrix}\frac{\partial u}{\partial x} - ia\begin{pmatrix} 0 & 1 \\ 1 & 0 \end{pmatrix} u = 0 \tag{105}$$

and the discriminant curve $D(\lambda,\mu) = 0$ is

$$\lambda^2 - c^2\mu^2 = a^2 \tag{106}$$

Components of the vector $u = \begin{pmatrix} u_1 \\ u_2 \end{pmatrix}$ satisfy the same scalar "string" equation

$$\frac{\partial^2 u}{\partial t^2} - c^2 \frac{\partial^2 u}{\partial x^2} + a^2 u = 0 \tag{107}$$

(In the case of Klein-Gordon equation $a = \frac{mc^2}{\hbar}$)

2) More general, let

$$X_0 = (K, H_0, \Phi_0, E_0, \sigma_0) \tag{108}$$

be a single operator colligation:

$$\frac{1}{i}(K-K^*) = \Phi_0^* \sigma_0 \Phi_0 \tag{109}$$

such that $\sigma_0 > 0$, and $\Phi_0(H_0) = E_0$. Let α_1, α_2 be selfadjoint matrices in C_m and let us define operators

$$A = \alpha_1 \otimes K + \alpha_2 \otimes I_0, \quad B = 1 \otimes K \tag{110}$$

in the tensor product $H = C_m \otimes H_0$, where I_0 and 1 are identity operators in H_0 and C_m respectively. It is evident that $AB = BA$ and

$$\begin{aligned} 2 \operatorname{Im}(A) &= \alpha_1 \otimes 2 \operatorname{Im}(K_0), \quad 2 \operatorname{Im}(B) = 1 \otimes 2 \operatorname{Im}(K_0) \\ 2 \operatorname{Im}(AB^*) &= 2 \operatorname{Im}(B^*A) = -\alpha_2 \otimes 2 \operatorname{Im}(K_0) \end{aligned} \tag{111}$$

Using (109) we can embed the pair (A,B) in the colligation

$$X = (A, B; C_m \otimes H_0, \Phi = 1 \otimes \Phi_0, E = C_m \otimes E_0, \sigma(A), \sigma(B)) \tag{112}$$

where

$$\sigma(A) = \alpha_1 \otimes \sigma_0, \quad \sigma(B) = 1 \otimes \sigma_0, \quad \gamma^{in} = \gamma^{out} = -\alpha_2 \otimes \sigma_0 \tag{113}$$

The discriminant polynomial is:

$$F(\lambda, \mu) = \det\left[\lambda \sigma(B) - \mu \sigma(A) + \gamma\right] = \left[\det(\lambda 1 - \mu \alpha_1 - \alpha_2)\right]^{\dim(E_0)} \cdot (\det \sigma_0)^m \tag{114}$$

In the particular case $\dim(E_0) = 1$, $\sigma_0 = 1$ we obtain:

$$D(\lambda, \mu) = \det(\lambda \cdot 1 - \mu \alpha_1 - \alpha_2) \tag{115}$$

and $\sigma(A) = \alpha_1$, $\sigma(B) = 1$, $\gamma = -\alpha_2$ \hfill (116)

Formulas (116) imply the following

Proposition. Let two aribtrary selfadjoint matrices α and β in C_m be given. There exist infinite many strict commutative colligations such that $\sigma(A) = \alpha$, $\sigma(B) = 1$, $\gamma^{in} = \gamma^{out} = \beta$. Indeed, to obtain such colligations we use the construction (112) with $\alpha_1 = \alpha$, $\alpha_2 = -\beta$.

§6. Couplings and Resolutions of Commutative Colligations

Let

$$X = (A, B; H, \Phi, E; \sigma(A), \sigma(B), \gamma^{in}, \gamma^{out}) \tag{117}$$

be a commutative regular colligation and H'' be a joint invariant subspace of A and B. It is easy to see that relations

$$P'AP'' = P''A^*P' = 0, \qquad P'BP'' = P''B^*P' = 0$$

$$A = A'P'+A''P''+i\Phi''\sigma(A)\Phi'P', \quad B = B'P'+B''P''+i\Phi''\sigma(B)\Phi'P' \tag{118}$$

are valid, where $H' = H \ominus H''$; $A' = P'A|_{H'}$, $B' = P'B|_{H'}$, $A'' = A|_{H''}$, $B'' = B|_{H''}$.

<u>Proposition</u>. Colligations

$$X' = (A',B';H',\Phi',E;\sigma(A),\sigma(B),\gamma'^{in},\gamma'^{out})$$
$$X'' = (A'',B'',H'',\Phi'',E;\sigma(A),\sigma(B),\gamma''^{in},\gamma''^{out}) \tag{119}$$

where

$$\Phi' = \Phi|_{H'}, \qquad \Phi'' = \Phi|_{H''}, \tag{120}$$

$$\gamma'^{in} = \gamma^{in}, \; \gamma'^{out} = \gamma^{in}+i[\sigma(A)\Phi'(\Phi')^*\sigma(B)-\sigma(B)\Phi'(\Phi')^*\sigma(A)] \tag{121}$$

$$\gamma''^{in} = \gamma''^{out} - [\sigma(A)\Phi''(\Phi'')^*\sigma(B)-\sigma(B)\Phi''(\Phi'')^*\sigma(A)] \tag{122}$$

are commutative regular colligations.

Indeed, condition (29):

$$\sigma(A)\Phi B^* - \sigma(B)\Phi A^* = \gamma^{in}\Phi \tag{123}$$

and relations $A^*|_{H'} = (A')^*$, $P''A^*|_{H'} = 0$, $B^*|_{H'} = (B')^*$, $P''B^*|_{H'} = 0$ imply

$$\sigma(A)\Phi'B'^* - \sigma(B)\Phi'A'^* = \gamma^{in}\Phi' \tag{124}$$

Anologously,

$$\sigma(A)\Phi''B'' - \sigma(B)\Phi''A'' = \gamma^{out}\Phi''$$

Now we define γ'^{out} and γ''^{in} by formulas (121), (122).

The colligations X', X'' of the form (119), which satisfy conditions (120), (121), (122) are said to be *the projections* of X onto H', H'' respectively.

<u>Remark</u>. From relations (48), (121), (122) and relation

$$\Phi\Phi^* = \Phi'\Phi'^* + \Phi''\Phi''^* \tag{125}$$

we conclude that

$$\gamma''^{in} = \gamma'^{out}$$

<u>The Matching Theorem</u>. The coupling $X = X' \vee X''$ of two given commutative regular colligations of the form (119) where $H = H' \oplus H''$ and (A,B) defined by (118), $\gamma^{in} = \gamma'^{in}$, $\gamma^{out} = \gamma''^{out}$ is a commutative regular colligation iff the condition

$$\gamma''^{in} = \gamma'^{out} \tag{126}$$

is valid. We will call this important equality the *matching* condition. In the case $\dim H = N < \infty$ there exists a chain $H = H_0 \supset H_1 \supset H_2 \supset \ldots \supset H_N = \{0\}$ of joint invariant subspaces of (A,B) such that $\dim(H_{k-1} \ominus H_k) = 1$. Then the corresponding projections on $H_k^\perp = H_{k-1} \ominus H_k$ are commutative regular colligations X_k with one-dimensional internal spaces $H_k^\perp$. To the coupling of colligations there corresponds the coupling of collective motions: the output of each link coincides with the input of the next link. Hence, we obtain a resolution of collective motions in a chain of elementary links with one-dimensional internal spaces. Conversely we can construct complicated systems using couplings of elementary systems. For applications of those chains to the so called determinantal representations of algebraic curves see [12]

Elementary Colligations

Let $M_0(\lambda_0, \mu_0)$ be a given point in C_2 and $\dim H = 1$. Let us define two operators in H:

$$Ah = \lambda_0 h, \qquad Bh = \mu_0 h \tag{127}$$

The following question arises: when it is possible to embed these operators in a regular colligation

$$X = (\lambda_0, \mu_0; H, \Phi, E; \sigma_1, \sigma_2, \gamma^{in}, \gamma^{out}) \tag{128}$$

with given E, $\sigma_1 = \sigma_1^*$, $\sigma_2 = \sigma_2^*$, $\gamma^{in} = (\gamma^{in})^*$. Conditions of regularity, written in the form

$$\sigma(A)\Phi B^* - \sigma(B)\Phi A^* = \gamma^{in}\Phi$$
$$\gamma^{out} = \gamma^{in} + i(\sigma(A)\Phi\Phi^*\sigma(B) - \sigma(B)\Phi\Phi^*\sigma(A)) \tag{129}$$
$$\sigma(A)\Phi B - \sigma(B)\Phi A = \gamma^{out}\Phi$$

imply

$$(\lambda_0^* \sigma_2 - \mu_0^* \sigma_1 + \gamma^{in})\Phi = 0, \tag{130}$$

$$\gamma^{out} = \gamma^{in} + i(\sigma_1 \Phi\Phi^*\sigma_2 - \sigma_2 \Phi\Phi^*\sigma_1), \tag{131}$$

$$(\lambda_0 \sigma_2 - \mu_0 \sigma_1 + \gamma^{out})\Phi = 0 \tag{132}$$

and conditions (23) imply

$$\frac{1}{i}(\lambda_0 - \lambda_0^*) = \Phi^*\sigma_1\Phi, \qquad \frac{1}{i}(\mu_0 - \mu_0^*) = \Phi^*\sigma_2\Phi \tag{132'}$$

From (130) we conclude that

$$(\text{Im } \lambda_0)\Phi^*\sigma_2\Phi = (\text{Im } \mu_0)\Phi^*\sigma_1\Phi$$

If $\text{Im } \mu_0 \neq 0$ and $\Phi^*\sigma_2\Phi \neq 0$ and $\text{sign}(\Phi^*\sigma_2\Phi) = \text{sign}(\text{Im } \mu_0)$ then multiplying by a scalar factor we obtain the normalization conditions (132'). Equation (130) or (132) have a nontrivial solution $\Phi \neq 0$ iff the point $M_0(\lambda_0,\mu_0)$ belongs to the discriminant curve :

$$D(\lambda_0,\mu_0) = \det(\lambda_0\sigma_2-\mu_0\sigma_1+\gamma^{out}) = \det(\lambda_0\sigma_2-\mu_0\sigma_1+\gamma^{in}) = 0$$

In particular, the embedding is always possible in the case $\sigma(B) > 0$ if $\text{Im } \mu_0 > 0$.

The characteristic functions

$$S_A(\lambda) = I - i(\lambda_0-\lambda)^{-1}\Phi\Phi^*\sigma_1 \;,\quad (\lambda \neq \lambda_0,\; \lambda \neq \lambda_0^*)$$
$$S_B(\mu) = I - i(\mu_0-\mu)^{-1}\Phi\Phi^*\sigma_2 \;,\quad (\mu \neq \mu_0,\; \mu \neq \mu_0^*) \tag{133}$$

map the $\text{Ker}(\lambda\sigma_2-\mu\sigma_1+\gamma^{in})$ onto the $\text{Ker}(\lambda\sigma_2-\mu\sigma_1+\gamma^{out})$ and the inverse operators have the form

$$S_A^{-1}(\lambda) = I + i(\lambda_0^*-\lambda)^{-1}\Phi\Phi^*\sigma_1$$
$$S_B^{-1}(\mu) = I + i(\mu_0^*-\mu)^{-1}\Phi\Phi^*\sigma_2 \tag{134}$$

If $X = (A,B;H,\Phi,E;\sigma(A),\sigma(B);\gamma^{in},\gamma^{out})$ is a regular colligation and $\dim H = N$ then using the resolution in a chain of elementary colligations we obtain the following factorizations of characteristic functions

$$S_A(\lambda) = \prod_{k=1}^{m} (I - i(\lambda_k-\lambda)^{-1}\Phi_k\Phi_k^*\sigma(A)) \;,$$
$$S_B(\mu) = \prod_{k=1}^{m} (I - i(\mu_k-\mu)^{-1}\Phi_k\Phi_k^*\sigma(B)) \;, \tag{135}$$

where (λ_k,μ_k) satisfy the equation

$$D(\lambda_k,\mu_k) = 0 \;,\quad k = 1,2,\ldots,N \tag{136}$$

and $\Phi^{(k)}$ can be found from recurrent formulas

$$\gamma^{(1)} = \gamma^{in} \;,$$
$$(\lambda_k^*\sigma(B) - \mu_k^*\sigma(A) + \gamma^k)\Phi_k = 0 \;,\quad k = 1,2,\ldots,N \tag{137}$$
$$\gamma^{(k+1)} = \gamma^{(k)} + i(\sigma(A)\Phi_k\Phi_k^*\sigma(B) - \sigma(B)\Phi_k\Phi_k^*\sigma(A)) \tag{138}$$
$$2\,\text{Im}(\mu_k) = \Phi_k^*\sigma(B)\Phi_k \tag{139}$$

All these equations have nontrivial solutions Φ_k according to the determinant conservation law.

§7. A One-Dimensional Wave Equation

Now we are going to solve the following problem: Given a wave equation

$$\frac{\partial^2 y}{\partial t^2} - c^2 \frac{\partial^2 y}{\partial x^2} + a^2 y = 0, \qquad (140)$$

can it be considered as a peripheral manifestation of a collective spatio-temporal motions? How does one find all such motions. This equation can be written in the form

$$(\frac{\partial}{\partial t} - c \frac{\partial}{\partial x})(\frac{\partial}{\partial t} + c \frac{\partial}{\partial x})y + a^2 y = 0 \qquad (141)$$

Introducing a vector $u = \binom{y}{z}$, where

$$z = ia^{-1}\theta^{-1}(\frac{\partial}{\partial t} - c \frac{\partial}{\partial x})y, \qquad |\theta| = 1, \qquad (142)$$

we obtain the following system of PDE:

$$\frac{a}{c}\begin{pmatrix}1 & 0\\ 0 & 1\end{pmatrix}\frac{\partial u}{\partial t} = a\begin{pmatrix}1 & 0\\ 0 & -1\end{pmatrix}\frac{\partial u}{\partial x} - i\frac{a^2}{c}\begin{pmatrix}0 & \theta\\ \theta^* & 0\end{pmatrix}u \qquad (143)$$

(In the case of Klein-Gordon equation $a = \frac{mc^2}{\hbar}$).
Comparing equations (43) and (143) we obtain

$$\sigma(A) = a\begin{pmatrix}1 & 0\\ 0 & -1\end{pmatrix}, \quad \sigma(B) = \frac{a}{c}\begin{pmatrix}1 & 0\\ 0 & 1\end{pmatrix}, \quad \gamma^{in} = \frac{a^2}{c}\begin{pmatrix}0 & \theta\\ \theta^* & 0\end{pmatrix} \qquad (144)$$

Our goal is to find all collective motions, corresponding to PDE (143), or, in other words, to find all regular commutative colligations, related to given $\sigma(A)$, $\sigma(B)$, γ^{in}. We use the generalized Cayley-Hamilton Theorem.

$$D(A,B)|_{\hat{H}} = 0, \qquad (145)$$

where $\hat{H}$ is the principal subspace (93) of X. This subspace reduces the operators A, B and they are selfadjoint on the orthogonal complement $H \ominus \hat{H}$. One can assume without restriction of generality that $H = \hat{H}$. The discriminant polynomial (53) in the case when $\sigma(A)$, $\sigma(B), \gamma^{in}$ are defined by (144) is

$$D(\lambda,\mu) = \frac{a^2}{c^2}(\lambda^2 - c^2\mu^2 - a^2) \qquad (146)$$

and the discriminant curve Γ is $\lambda^2 = c^2\mu^2 + a^2$. Relations (145), (146) imply

$$A^2 - c^2 B^2 = a^2 I \qquad (147)$$

Hence

$$\frac{A-cB}{a} \cdot \frac{A+cB}{a} = I \tag{148}$$

Let us introduce operators

$$Q = a^{-1}(A+cB) , \quad Q^{-1} = a^{-1}(A-cB) \tag{149}$$

Formulas (144) imply

$$\sigma(Q) = 2\begin{pmatrix} 1 & 0 \\ 0 & 0 \end{pmatrix} , \quad \sigma(Q^{-1}) = -2\begin{pmatrix} 0 & 0 \\ 0 & 1 \end{pmatrix} \tag{150}$$

Hence the range of $Q-Q^*$ is one-dimensional and $2\,\mathrm{Im}Q$ has the following form

$$\frac{1}{i}(Q-Q^*)f = 2(f,g_0)g_0 , \quad (g_0 \in H) \tag{151}$$

Let $X = (A,B;H,\Phi,E;\sigma(A),\sigma(B))$ be a colligation and a_k ($k=1,2,\ldots,n$) be an orthonormal basis in E. The elements $g_k = \Phi^* a_k$ are said to be *channel elements* of the colligation. It is easy to see that

$$(2\,\mathrm{Im}\,A)f = \sum_{k,j=1}^{n} g_k \sigma_{kj}(A)(f,g_j) \; ; \quad (2\,\mathrm{Im}\,B)f = \sum_{k,j=1}^{n} g_k \sigma_{kj}(B)(f,g_j), \tag{152}$$

where $\sigma_{kj} = (\sigma a_j, a_k)$.

There are analogous formulas for $2\mathrm{Im}(AB^*)$, $2\mathrm{Im}(B^*A)$. If $\sigma(A)$, $\sigma(B)$, γ^{in} are defined by formulas (144) then relations (149) imply

$$A = \frac{a}{2}(Q+Q^{-1}) , \quad B = \frac{a}{2c}(Q-Q^{-1}) \tag{153}$$

and the relation (151) implies

$$\frac{1}{i}[Q^{-1}-(Q^*)^{-1}]f = iQ^{-1}(Q-Q^*)(Q^*)^{-1}f = -2(f,Q^{-1}g_0)Q^{-1}g_0 \tag{154}$$

Hence

$$\frac{1}{i}(A-A^*)f = a[(f,g_0)g_0 - (f,Q^{-1}g_0)Q^{-1}g_0] ,$$

$$\frac{1}{i}(B-B^*)f = \frac{a}{c}[(f,g_0)g_0 + (f,Q^{-1}g_0)Q^{-1}g_0] ,$$

$$\frac{1}{i}(AB^*-BA^*) = -\frac{a^2}{c}[(f,g_0)Q^{-1}g_0 + (f,Q^{-1}g_0)g_0]$$

Taking $E = C_2$, $a_1 = \binom{1}{0}$, $a_2 = \binom{0}{1}$ we obtain that

$$g_1 = \theta^* g_0 , \quad g_2 = -Q^{-1}g_0 , \quad (|\theta| = 1) \tag{155}$$

are two channel elements which satisfy the colligation conditions (152) with $\sigma(A)$, $\sigma(B)$, γ^{in}, defined by (144). If g_0 and $Q^{-1}g_0$ are linearly independent then $\dim G = 2$ and the colligation is strict. If g_0, $Q^{-1}g_0$ are linearly dependent then

$H = \text{span}\{A^k B^j \phi^*(E)\}$ is one-dimensional. In this case one can take $H_0 = C_1$ and

$$Q = \zeta_0, \quad A = \lambda_0 = \frac{a}{2}(\zeta_0 + \zeta_0^{-1}), \quad B = \mu_0 = \frac{a}{2c}(\zeta_0 - \zeta_0^{-1}) \quad (156)$$

where ζ_0 is a complex number such that $\text{Im } \zeta_0 > 0$. Taking

$$g_1 = \theta g_0, \quad g_2 = -\zeta_0^{-1} g_0, \quad g_0 = (2 \text{ Im } \zeta_0)^{\frac{1}{2}} \quad (157)$$

one can easily verify the regularity conditions (29), (30).

<u>Calculation of γ^{out}</u>: We use the formula

$$\gamma^{out} = \gamma^{in} + i[\sigma(A)\Phi\Phi^*\sigma(B) - \sigma(B)\Phi\Phi^*\sigma(A)],$$

where

$$\Phi\Phi^* = \begin{pmatrix} (g_1, g_1), & (g_2, g_1) \\ (g_1, g_2), & (g_2, g_2) \end{pmatrix}$$

Then, using (155) we obtain

$$\gamma^{out} = \frac{a^2}{c} \begin{pmatrix} 0 & \theta w(0) \\ \theta^* w^*(0), & 0 \end{pmatrix} \quad (158)$$

where

$$w(z) = 1 - 2i((Q - zI)^{-1} g_0, g_0) \quad (159)$$

is the characteristic function of the colligation

$$X_0 = (Q; H, \Phi_0, E_0 = C_1, \sigma_0) \quad (160)$$

and

$$\Phi_0 f = (f, g_0), \quad \Phi_0^*(1) = g_0, \quad \sigma_0 = 2 \quad (161)$$

Now one can find $E^{in}(M)$, $(M \in \Gamma)$: if $\hat{u} = \begin{pmatrix} u_1 \\ u_2 \end{pmatrix} \in E^{in}(M)$ then

$$(\lambda - c\mu)u_1 + a\theta u_2 = 0,$$
$$a\theta u_1 + (\lambda + c\mu)u_2 = 0, \quad (\lambda^2 - c^2\mu^2 = a^2)$$

Hence

$$\hat{u} = \text{const.} \begin{pmatrix} z \\ -\theta^* \end{pmatrix}, \quad z = a^{-1}(\lambda + c\mu) \quad (162)$$

Analogously,

$$v = \text{const.} \begin{pmatrix} z \\ -\theta^* w^*(0) \end{pmatrix}, \quad \hat{v} \in E^{out}(M) \quad (163)$$

<u>Remark</u>. Equations of collective motions can be written in the form

$$ia^{-1}(\tfrac{\partial}{\partial t} + c\tfrac{\partial}{\partial x})f + Qf = \Phi^*\sigma(Q)(u)$$

$$ia^{-1}(\tfrac{\partial}{\partial t} - c\tfrac{\partial}{\partial x})f + Q^{-1}f = \Phi^*\sigma(Q^{-1})(u) \qquad (164)$$

$$v = u - i\Phi(f)$$

Introducing new variables

$$\xi = \tfrac{a}{2}(t + \tfrac{x}{c}), \qquad \eta = \tfrac{a}{2}(t - \tfrac{x}{c}) \qquad (165)$$

we obtain

$$i\tfrac{\partial f}{\partial \xi} + Qf = \Phi^*\sigma(Q)(u) \qquad (166)$$

$$i\tfrac{\partial f}{\partial \eta} + Q^{-1}f = \Phi^*\sigma(Q^{-1})(u) \qquad (167)$$

The equation (166) corresponds to the "section" of collective motions, related to the motion $x = ct + x_0$ along the x-axis and the equation (167) corresponds to the motion $x = -ct + x_0$ in the opposite direction.

Let us consider a collective plane wave

$$(u,f,v) = (\hat{u},\hat{f},\hat{v})e^{i(\lambda t + \mu x)}, \qquad (M(\lambda,\mu) \in \Gamma) \qquad (168)$$

In variables ξ, η the exponential factor has the form

$$e^{i(z\xi + z^{-1}\eta)} \qquad (169)$$

where

$$z = a^{-1}(\lambda + \mu c), \qquad z^{-1} = a^{-1}(\lambda - \mu c) \qquad (170)$$

and

$$\lambda = \tfrac{a}{2}(z + z^{-1}), \qquad \mu = \tfrac{a}{2c}(z - z^{-1}) \qquad (171)$$

is a parametric representation of the curve $\lambda^2 - c^2\mu^2 = a^2$. The JCF maps $E^{in}(M)$ into $E^{out}(M)$:

$$\hat{S}(M)\hat{u} = p(z)\hat{v}, \qquad (\hat{u} \in E^{in}(M), \hat{v} \in E^{out}(M)), \qquad (172)$$

where $p(z)$ is a scalar. To find $p(z)$ we use the definition (65) of JCF:

$$\hat{S}(M)\hat{u} = I - i\Phi(\alpha A + \beta B - (\alpha\lambda + \beta\mu)I)^{-1}\Phi^*\sigma(\alpha A + \beta B)\hat{u}$$

Putting $\alpha = a^{-1}, \beta = a^{-1}c$ we obtain

$$\hat{S}(M)\hat{u} = [I - i\Phi(Q - zI)^{-1}\Phi^*\sigma(Q)]\hat{u}, \qquad (173)$$

where $\sigma(O) = 2\begin{pmatrix} 1 & 0 \\ 0 & 0 \end{pmatrix}$. In the matrix form

$$S(M)u = \left[I - 2i \begin{pmatrix} (R_z g_1, g_1), & 0 \\ (R_z g_1, g_2), & 0 \end{pmatrix} \right] u, \tag{174}$$

where $R_z = (Q-zI)^{-1}$, $g_1 = \theta g_0$, $g_2 = -Q^{-1} g_0$. Relations (172), (174) imply $p(z) = (z)$, where $w(z)$ is the characteristic function of colligation (160). Thus, for JCF the following formula

$$S(M) \begin{pmatrix} z \\ -\theta * \end{pmatrix} = w(z) \begin{pmatrix} z \\ -[\theta w(0)]* \end{pmatrix} \tag{175}$$

is valid. Taking the spatial characteristic matrix we obtain

$$\hat{S}_B(\mu) \begin{pmatrix} \theta z \\ -1 \end{pmatrix} = w(z) \begin{pmatrix} \theta z \\ -w*(0) \end{pmatrix}, \quad S_B(\mu) \begin{pmatrix} \theta \\ z \end{pmatrix} = w(-z^{-1}) \begin{pmatrix} \theta \\ zw*(0) \end{pmatrix} \tag{176}$$

Now we can write the matrix $S_B(\mu)$ in the basis $a_1 = \binom{1}{0}$, $a_2 = \binom{0}{1}$. After calculations we obtain

$$S_B(\omega) = \begin{pmatrix} 1 & 0 \\ 0 & w*(0) \end{pmatrix} V^{-1} \begin{pmatrix} w(-z^{-1}), & 0 \\ 0, & w(z) \end{pmatrix} V, \tag{177}$$

where

$$V = (z^2-1)^{-\frac{1}{2}} \begin{pmatrix} 1 & z \\ -z, & 1 \end{pmatrix} \begin{pmatrix} 1 & 0 \\ 0 & \theta \end{pmatrix}, \quad (z\text{-real}) \tag{178}$$

If

$$u(t,x) = \hat{u}_1 e^{i(\mu x + \lambda t)} + \hat{u}_2 e^{i(\mu x - \lambda t)} \tag{179}$$

is an input, satisfying PDE (143), and

$$v(t,x) = \hat{v}_1 e^{i(\mu x + \lambda t)} + \hat{v}_2 e^{i(\mu x - \lambda t)} \tag{180}$$

the corresponding output then $v = S_B(\mu)(u)$.
Analogously,

$$S_A(\lambda) = \begin{pmatrix} 1 & 0 \\ 0 & w*(0) \end{pmatrix} V_1^{-1} \begin{pmatrix} w(z), & 0 \\ 0 & w(z^{-1}) \end{pmatrix} V_1, \tag{181}$$

where

$$V_1 = (z^2-1)^{-\frac{1}{2}} \begin{pmatrix} z & -1 \\ -1, & z \end{pmatrix} \begin{pmatrix} 1 & 0 \\ 0 & \theta \end{pmatrix}$$

If

$$u(t,x) = \hat{u}_1 e^{i(\lambda t + \mu x)} + \hat{u}_2 e^{i(\lambda t - \mu x)} \tag{182}$$

is an input, satisfying PDE (143) and

$$\tilde{v}(t,x) = \tilde{v}_1 e^{i(\lambda t+\mu x)} + \tilde{v}_2 e^{i(\lambda t-\mu x)} \tag{183}$$

the corresponding output then $\tilde{v} = S_A(\lambda)(\tilde{u})$.

§8. Symmetries

Symmetry I: Let us consider the transformation $\lambda \to -\lambda$, $\mu \to \mu$. Relations

$$\lambda = \frac{a}{2}(z+z^{-1}), \qquad \mu = \frac{a}{2c}(z-z^{-1}) \tag{184}$$

imply that this transformation is equivalent to the transformation $z \to z^{-1}$. The symmetry condition in this case is

$$w(-z^{-1}) = w(z) \tag{185}$$

Then $w(0) = w(\infty) = 1$ and, taking in account (177), we obtain

$$S_B(\mu) = w(z)\begin{pmatrix} 1 & 0 \\ 0 & 1 \end{pmatrix} \tag{186}$$

Let us consider two special cases:

a) If H is finite dimensional then $w(z)$ is a rational function with poles z_k, z_k^{-1} ($k = 1, 2, \ldots, m$):

$$w(z) = \prod_{k=1}^{m}\left(\frac{z-z_k^*}{z-z_k} \cdot \frac{z+(z_k^*)^{-1}}{z+z_k^{-1}}\right) = \prod_{k=1}^{m} \frac{\mu-\mu_k^*}{\mu-\mu_k} \tag{187}$$

where

$$\mu_k = \frac{a}{2c}(z_k - z_k^{-1}) \tag{188}$$

b) If $w(z)$ has transcendental singularities z_0, $-z_0^{-1}$ they must be real numbers and

$$w(z) = \exp\left(\frac{i\ell}{\mu-\mu_0}\right), \qquad (\ell > 0), \tag{189}$$

where $\mu_0 = \frac{a}{2c}(z_0 - z_0^{-1})$.

In general case the condition $w(z) = w(-z^{-1})$ implies that $w(z)$ is a single-valued analytic function of μ: $w(z) = W(\mu)$.

From Theorem 1 we conclude that $w(z)$ belongs to the class $\Omega(\sigma_0)$, where $\sigma_0 = 2$, $\dim E_0 = 1$. Hence, $w(z)$ has the following properties:

1) $w(z)$ is a meromorphic function in the upper halfplane and $w(\infty) = 1$.
2) $w(z)$ is holomorphic in a neighborhood of $z=\infty$
3) $|w(z)| > 1$, $(\text{Im } z > 0)$
4) $|w(z)| = 1$, $(\text{Im } z = 0, |z| > N)$

Using the condition $w(z) = w(-z^{-1})$ and the transformation $\mu = \frac{a}{2c}(z-z^{-1})$ we conclude that $W(\mu) = w(z)$ satisfies conditions of type 1)-4) in the μ-plane. Hence, Theorem 3 implies that $W(\mu)$ is the characteristic function

$$W(\mu) = 1 - i\Phi'(K-\mu I)^{-1}\Phi'\sigma' \tag{190}$$

of a colligation of the form

$$X' = (K, H', \Phi', E', \sigma') \tag{191}$$

where $E' = C_1$, $\sigma'_0 = \frac{a}{c}$ and $H' = \text{span}\{K^n \Phi'(E')\}$. Then

$$\frac{1}{i}(K-K^*)f = \frac{a}{c}(f, g')g' \tag{191'}$$

Let us consider now the following composit colligation

$$X = (A, B; H, \Phi, E; \sigma(A), \sigma(B), \gamma^{in}, \gamma^{out}), \tag{192}$$

where

$$A = \begin{pmatrix} cK, & -aI \\ -aI, & -cK \end{pmatrix}, \quad B = \begin{pmatrix} K, & 0 \\ 0, & K \end{pmatrix}, \quad H = H' \oplus H', \quad E = C_2 \tag{193}$$

$$\sigma(A) = a\begin{pmatrix} 1 & 0 \\ 0 & -1 \end{pmatrix}, \quad \sigma(B) = \frac{a}{c}\begin{pmatrix} 1 & 0 \\ 0 & 1 \end{pmatrix}, \quad \Phi\begin{pmatrix} f_1 \\ f_2 \end{pmatrix} = \begin{pmatrix} (f_1, g') \\ (f_2, g') \end{pmatrix},$$

$$\Phi^*\begin{pmatrix} u_1 \\ u_2 \end{pmatrix} = \begin{pmatrix} u_1 g' \\ u_2 g' \end{pmatrix} \tag{194}$$

It is easy to check that

$$2\,\text{Im}(AB^*) = 2\,\text{Im}(B^*A) = \frac{a^2}{c}\begin{pmatrix} (f_2, g')g' \\ (f_1, g')g' \end{pmatrix}$$

Hence, X is a strict commutative colligation such that

$$\gamma^{in} = \gamma^{out} = \frac{a^2}{c}\begin{pmatrix} 0 & 1 \\ 1 & 0 \end{pmatrix}.$$

It is evident that spatial CF of X is

$$S_B(\mu) = W(\mu)\begin{pmatrix} 1 & 0 \\ 0 & 1 \end{pmatrix} \tag{195}$$

where

$$W(\mu) = 1 - i\frac{a}{c}((K-\mu I)^{-1}g', g') \tag{196}$$

is the characteristic function (190). From relations (195), (186) it follows that CF (195) coincides with the CF (186). Using the Theorem 6 and its corollary we obtain the following.

Proposition. If a collective spatio-temporal system, related to the wave equation (140), is invariant under transformation $\lambda \to -\lambda$, $\mu \to \mu$

then the corresponding colligation is unitary equivalent to the composit colligation, which defined by formulas (193), (194).

<u>Symmetry</u> II. Let us consider now the transformation $\lambda \to -\lambda^*$, $\mu \to -\mu^*$. The symmetry condition in this case is

$$w(z) = w^*(z^*) \qquad (197)$$

If dim H < ∞ then $w(z)$ is a rational function with poles z_k, $-z_k^*$ (k = 1,2,...,m).

If the system is symmetric with respect to both transformations:

$$w(-z) = [w(x^*)]^* = w(z^{-1}) \qquad (198)$$

then $w(z) = W(\mu)$ and

$$W(-\mu) = W^*(\mu^*) \qquad (199)$$

<u>Resonances</u>. We assume now that the system under consideration is symmetric with respect to the transformation $\lambda \to -\lambda$, $\mu \to \mu$. We can use in this case formulas (193). If dim(H') = m < ∞ then the characteristic function $W(\mu)$ is given by formula (187):

$$W(\mu) = \prod_{k=1}^{m} \frac{\mu - \mu_k^*}{\mu - \mu_k} \qquad (200)$$

and the operator K can be reduced to the triangular form (18):

$$f_k = \mu_k f_k + i \sum_{j=k+1}^{m} \rho_k \rho_j f_j, \quad (k = 1,2,\ldots,m), \qquad (201)$$

where $\rho_j = (2 \operatorname{Im} \mu_j)^{1/2}$. It is easy to check that

$$\frac{1}{i}(K - K^*)f = (f,g)g, \qquad (202)$$

where

$$g_k = \rho_k, \quad (k = 1,2,\ldots,m). \qquad (203)$$

If we consider a collective plane wave of the form (168) then the internal state can be obtained from formulas (63), where λ and μ are connected by the relation $\lambda^2 - c^2\mu^2 = a^2$. Choosing an appropriate constant factor in formulas (163) we obtain

$$f_0 = \frac{a}{c}(1+z^2)^{-1/2} \begin{pmatrix} (K-\mu I)^{-1}g', & 0 \\ 0, & (K-I)^{-1}g' \end{pmatrix} \begin{pmatrix} \theta z \\ -1 \end{pmatrix}, \qquad (204)$$

where g' is the channel element (191'):

$$g'_k = \left(\frac{c}{a}\right)^{1/2} \rho_k, \quad (k = 1,2,\ldots,m) \qquad (205)$$

If m=1 then

$$f_0 = (1+z^2)^{-\frac{1}{2}}(2\frac{a}{c}\text{Im }\mu_0)^{\frac{1}{2}}\begin{pmatrix}z\theta\\-1\end{pmatrix}(\mu_0-\mu)^{-1}, \tag{206}$$

$$|f_0|^2 = 2\frac{a}{c}\frac{\text{Im }\mu_0}{(\mu-\text{Re }\mu_0)^2+(\text{Im }\mu_0)^2}, \tag{207}$$

where $z = a^{-1}(\lambda+c\mu)$. Thus, $\mu = \text{Re }\mu_0$ is the resonance value.

A singularity μ_0 on the real axis. In this case we use formula (189) and the corresponding triangular model is:

$$(Kf)(\xi) = \mu_0 f(\xi) + i\int_0^\xi f(\eta)d\eta, \qquad (0 \leq \xi \leq \ell) \tag{208}$$

The corresponding channel element is

$$g'(\xi) \equiv (\frac{c}{a})^{\frac{1}{2}}, \qquad (0 \leq \xi \leq \ell) \tag{209}$$

Using (204) we obtain

$$f_0 = (\frac{a}{c})^{\frac{1}{2}}(1+z^2)^{-\frac{1}{2}}\begin{pmatrix}z\theta\\-1\end{pmatrix}(K-\mu I)^{-1}(1) \tag{210}$$

Denoting $h(\xi) = (K-\mu I)^{-1}(1)$ we obtain

$$(\mu_0-\mu)h(\xi) + i\int_0^\xi h(\eta)d\eta = 1, \qquad 0 \leq \xi \leq \ell \tag{211}$$

Hence,

$$(\mu_0-\mu)\frac{dh}{d\xi} + ih(\xi) = 0 \tag{212}$$

and $h(0) = (\mu_0-\mu)^{-1}$.

Then

$$h(\xi) = (\mu_0-\mu)^{-1}\exp\left(\frac{i\xi}{\mu-\mu_0}\right) \tag{213}$$

Hence

$$|f_0|^2 = \frac{a}{c}\int_0^\ell |h(\xi)|^2 d\xi = \frac{a\ell}{c}(\mu-\mu_0)^{-2} \tag{214}$$

Comparing formulas (207) and (214) we conclude that if $\text{Im }\mu_0 \to 0$ in (207) then $|f_0|^2$ behave like $\delta(\mu-\mu_0)$. In spite of this $|f_0|^2$ in the case (214) behave like $(\mu-\mu_0)^{-2}$, ($\text{Im }\mu_0 = 0$) and $\mu = \mu_0$ corresponds to a "*giant*" resonance.

Collective Green Functions

Let X be a regular commutative colligation such that $\sigma(B) > 0$

$\lim\limits_{x \to +\infty} e^{ixB} f = 0$, $(f \in H)$. Then, according to §4, the transformation

$$F(\tau,\xi) = \Phi \ell^{i(\tau A+\xi B)} f \tag{215}$$

defines an isometric mapping of H onto a subspace $\tilde{H}$ of solutions of the output PDE. In this representation the colligation X has the following form

$$\tilde{X} = \left(-i\frac{\partial}{\partial \tau}, -i\frac{\partial}{\partial \xi}; \tilde{H}, \tilde{\Phi}, E; \sigma(A), \sigma(B), \gamma^{in}, \gamma^{out}\right), \tag{216}$$

where

$$\tilde{\Phi}[F(\tau,\xi)] = F(0,0). \tag{217}$$

The transformation $\tilde{\Phi}$ maps $\tilde{H}$ into E:

$$\tilde{\Phi}[F(\tau,\xi)] = \sum_{k=1}^{n} (\tilde{\Phi}F, a_k) a_k = \sum_{k=1}^{n} < F, G_k > a_k, \tag{218}$$

where

$$G_k(\tau,\xi) = \tilde{\Phi}^*(a_k) \tag{219}$$

are the output representations of channel elements $g_k = \Phi^*(a_k)$. Comparing (228) and (227) we conclude that scalar products $< F, G_k >$ have the form:

$$< F, G_k > = F_k(0,0) = \int_0^\infty (\sigma(B)F(0,\xi), G_k(0,\xi)) d\xi, \tag{220}$$

where $F_k(\tau,\xi) = (F(\tau,\xi), a_k)$. Hence, functions $\sigma(B)G_k(0,x)$ in the space $\tilde{H}_0$ play the role of singular functions $\delta(x)$:

$$\int_0^\infty (F(0,x), \sigma(B)G_k(0,x)) dx = \int_0^\infty F_k(0,x)\delta(x) dx \tag{221}$$

<u>Definition</u>. The output representations $G_k(\tau,\xi)$ of channel elements g_k are said to be *collective Green* functions and the corresponding modes $G_k(0,\xi)$ are said to be *channel modes*.

Collective Green functions satisfy the output PDE:

$$\sigma(B)\frac{\partial G}{\partial \tau} = \sigma(A)\frac{\partial G}{\partial \xi} - i\gamma^{out} G.$$

The channel modes $G_k(0,\xi)$ play the role of δ-functions.

<u>Example</u>. Let us consider the colligation X defined by formulas (201) and assume $\mu_1 = \mu_2 = \ldots = \mu_m = \mu_0$. Then the corresponding characteristic function (200) is

$$W(\mu) = \left(\frac{\mu-\mu_0^*}{\mu-\mu_0}\right)^m \tag{222}$$

and the corresponding operator Kh is

$$h_k = \mu_0 h_k + i\rho_0^2 \sum_{j=k+1}^{m} h_j, \quad (k = 1,2,\ldots,m), \tag{223}$$

where $\rho_0 = (2 \operatorname{Im} \mu_0)^{1/2}$. The channel element (205) in this case is

$$g'(k) = \left(\frac{c}{a}\right)^{1/2} \rho_0$$

Relations (194) imply

$$g_1 = \Phi^*(a_1) = \begin{pmatrix} g' \\ 0 \end{pmatrix} \qquad g_2 = \Phi^*(a_2) = \begin{pmatrix} 0 \\ g' \end{pmatrix}$$

After calculations we obtain channel modes

$$G_1(0,x) = \begin{pmatrix} 1 \\ 0 \end{pmatrix} f(x), \qquad G_2(0,x) = \begin{pmatrix} 0 \\ 1 \end{pmatrix} f(x),$$

where

$$f(x) = \left(\frac{c}{a}\right)^{1/2} \rho_0 L_{n-1}^{(1)}(x\rho_0), \qquad 0 \leqslant x < \infty$$

and

$$L_k^{(1)}(x) = x^{-1} e^{-x} \frac{1}{k!} \frac{d^k}{dx^k} (x^{k+1} e^{-x}), \qquad k = 0,1,2,\ldots$$

are the Laguerre polynomials of order 1.

References

1. V.M. Agranovich, Optical properties of crystals in the excitonic region of the spectrum, in: "Optical Properties of Solids" (F. Abelès, Ed.), Chap. 6, Laboratoire d'Optique, Paris, France, 1972, North-Holland.
2. M.S. Brodskij, "Triangular and Jordan Representations of Operators", Amer. Math. Soc. Trans., Vol. 32, 1971.
3. Gohberg, I.C. and Krein, M.G., a) Introduction to the theory of linear nonselfadjoint operators in Hilbert space, Transl. Math. Mon. v. 18 (1969). b) Theory and Application of Volterra operators in Hilbert Space, Transl. Math. Mon. v. 18 (1970).
4. Douglas, R.G., Canonical Models, Math. Surveys, v. 13 (1974).
5. N. Kravitsky, Regular colligations for several operators in Banach space, Int. Equat. and Operator Theory, v. 3/1 (1983).
6. M.S. Livšic, Cayley-Hamilton Theorem, Int. Eq. and Operator Theory, v. 6 (1983).
7. M.S. Livšic, Collective Motions of Spatio-Temporal Systems, J. of Math. Analysis and Appl., Vol. 116, No. 1, 1968.

8. M.S. Livšic, Commuting nonselfadjoint operators and mappings of vector bundles on algebraic curves, Operator Theory: Advances and Appl., v. 19 (1986), (Proceedings Workshop Amsterdam, June 4-7, 1985, Birkhäuser.
9. M.S. Livšic, "Operators, Oscillations, Waves", Amer. Math. Soc. Transl., vol. 34, Providence, R.I., 1973.
10. B. Sz. Nagy and C. Foiaş, "Harmonic Analysis of Operators in Hilbert Space", North-Holland, Amsterdam, 1970.
11. N.K. Nikolskij, "Lectures on Shift Operators", Springer-Verlag, Berlin, 1985.
12. V. Vinnikov, Self-Adjoint determinantal representations of real irreducible cubics, Operator Theory. Advances and Appl., Vol. 19, (1986), Birkhäuser.
13. L.L. Waksman, Harmonic analysis of multiparameter semigroups of contractions, Lecture Notes (Part II of this volume)

This research was supported by the Fund for Basic Research Administrated by The Israel Academy of Sciences and Humanities.

The David and Helena Zlotowski chair in Operator Theory and Systems Theory. Department of Mathematics and Computer Sciences, Ben Gurion University of the Negev, Beer Sheva, Israel.

HARMONIC ANALYSIS OF MULTI-PARAMETER SEMIGROUPS OF CONTRACTIONS

Leonid Waksman

Translation by Tatiana de Branges

CONTENTS

INTRODUCTION.. 42
CHAPTER 1. MULTI-PARAMETER SEMIGROUPS OF ISOMETRIES
  IN HILBERT SPACE.. 47
  §1. Functional Models of Isometric Representations........ 47
  §2. Isometric Representations and Hyperbolic Systems
      of Equations.. 55
  §3. Isometric Representations of Finite Rank of
      Nonunitarity.. 59
      Supplement 1.. 64
      Supplement 2.. 66
CHAPTER 2. ON ISOMETRIC REPRESENTATIONS OF RANK OF
  NONUNITARITY THREE.................................... 70
CHAPTER 3. UNITARY DILATIONS OF MULTI-PARAMETER SEMIGROUPS
  OF CONTRACTIONS....................................... 81
  §1. The Functional Model of Unitary Dilations............. 81
  §2. Core Unitary Dilations of Contractive Representations. 84
CHAPTER 4. FUNCTIONAL MODELS OF MULTI-PARAMETER SEMIGROUPS
  OF CONTRACTIONS....................................... 90
  §1. Model Construction within Unitary Equivalence......... 90
  §2. Functional Models and Hardy Spaces on Riemann
      Surfaces.. 95
      Supplement 1... 101
APPENDIX. TRIANGULAR MODELS OF COMMUTING LINEAR OPERATORS
  AND NONLINEAR DIFFERENTIAL EQUATIONS.................. 104
REFERENCES 113
SUBJECT INDEX... 115

19 February 1983   L. L. Waksman

INTRODUCTION

The traditional approach to the spectral analysis of linear operators belonging to certain classes consists of the construction of a related functional calculus. To every operator A of the given class, there corresponds a representation
$$f(z) \to f(A)$$
of a given Banach algebra of functions, appropriate to all operators of that class. For example the von Neumann inequality [1]
$$\|f(A)\| \leq \sup_{\operatorname{Im} z > 0} |f(z)| \qquad (1)$$
holds for any bounded dissipative operator A in a Hilbert space,
$$\frac{1}{i}(A - A^*) \geq 0 .$$
Here $f(z)$ is any rational function which is bounded in the upper half-plane. Because of inequality (1), the mapping $f(z) \to f(A)$ extends by continuity to a representation of the algebra of functions which are holomorphic in the upper half-plane and have continuous extension to the closed half-plane.

The spectral analysis of dissipative operators and of related one-parameter semigroups of contractions $\exp(iAt)$ has led to a series of deep results within the context of the theory of unitary dilations and characteristic operator functions.

It is known that, in the case of a family of pair-wise commuting dissipative operators $A_1, \ldots, A_n$, an inequality of the form
$$\|f(A_1, \ldots, A_n)\| \leq \sup |f(z_1, \ldots, z_n)|$$
for all f, with the least upper bound taken over $\operatorname{Im} z_1 > 0, \ldots,$ $\operatorname{Im} z_n > 0$, holds only when $n = 2$.

Thus traditional methods in the theory of unitary dilations fail when $n \geq 3$. Moreover it is not clear for which Banach algebra a naturally related functional calculus is to be expected.

In a series of papers on operator representations of function algebras, the difficulties arising were more circumvented than overcome [2].

The recent investigations of M. S. Livšic indicate a direction for further research [20]. The present work arose from a number of stimulating conversations with M. S. Livšic. The author takes advantage of this opportunity to express to him the deepest gratitude.

Consider a pair of commuting operators A_1, A_2 in H. A subspace L is said to be a reducing subspace if it is invariant under A_1 and A_2 and if its orthogonal complement is invariant under A_1 and A_2. The pair A_1, A_2 is said to be completely nonselfadjoint if the only reducing subspace of H to which the restrictions of A_1 and A_2 are selfadjoint is the zero subspace.

Define G to be the closed span of the ranges of Im A_1 and Im A_2,
$$\text{Im } A = \frac{A - A^*}{2i}.$$
Assume that the dimension of G is finite. Consider the polynomial
$$\det(z_1, z_2) = \det_G (\frac{A_2 - A_2^*}{i} z_1 - \frac{A_1 - A_1^*}{i} z_2$$
$$+ \frac{1}{i} \{ (A_2^* - A_2) A_1 - (A_1^* - A_1) A_2 \})$$
where the subscript G indicates that the determinant is applied to the restriction of the transformation to G.

A generalization of the Cayley-Hamilton theorem which is due to M. S. Livšic states that $d(A_1, A_2) = 0$. This result means that the study of a pair of commuting operators leads to problems in the theory of functions on the algebraic curve consisting of those points (z_1, z_2) which satisfy the equation $d(z_1, z_2) = 0$, and not to a substantially more complicated problem in the theory of functions of two complex variables [3], as was previously believed.

The identity $d(A_1, A_2) = 0$ results from the fact that, for all f in H, the vector-function
$$F(t_1, t_2) = \exp(iA_1 t_1 + iA_2 t_2) f$$
is a solution of the "hyperbolic system of equations"
$$(\frac{A_2 - A_2^*}{i} \frac{1}{i} \frac{\partial}{\partial t_1} - \frac{A_1 - A_1^*}{i} \frac{1}{i} \frac{\partial}{\partial t_2} + \frac{A_2^* A_1 - A_1^* A_2}{i}) F = 0 . \qquad (2)$$

Indeed it is not difficult to determine a differential operator with operator coefficients which, when applied to each side of equation (2), gives the equation
$$d(\frac{1}{i} \frac{\partial}{\partial t_1}, \frac{1}{i} \frac{\partial}{\partial t_2}) P_G F = 0 .$$

Here P_G denotes the orthogonal projection of H onto the subspace G. The identity
$$P_G \exp(iA_1 t_1 + iA_2 t_2) d(A_1, A_2) h = 0$$
holds for every element h of H. It remains to make use of the fact that no nonzero element f of H exists such that
$$P_G \exp(iA_1 t_1 + iA_2 t_2) f = 0$$

identically since the pair A_1, A_2 is assumed to be completely nonselfadjoint. This completes the verification that
$$d(A_1, A_2) = 0 .$$
The following scheme for the study of contractive and dissipative operators is due to Nagy and Foias.

The first step is concerned with the study of the operator of "multiplication by z" in a space of vector functions. It is shown that its compression into "partially invariant" subspaces exhausts the whole class of operators of interest, when they are carefully examined from the point of unitary equivalence. In other words, a functional model of the given operator is constructed as the first step. Then the methods of function theory are applied to the model operator. In the present work the functional models of isometric and contractive representations of certain multi-parameter semigroups $K \subset R^n$, and the functional models of certain families of commuting dissipative operators, are obtained according to the same scheme.

The study of model operators in Hardy spaces on Riemann surfaces makes use of the description of common invariant subspaces for multiplication by holomorphic functions which is due to Hasumi, Voichick and Forelli [5,6]. In the Nagy-Foias theory, in the proof of the spectral mapping theorem, and in the problems of spectral analysis for the point spectrum, very delicate results in the "spectral" theory of functions on the circle are applied [18,19]. The generalization of these results to a Riemann surface [7,8,9,10,15] would clearly allow a detailed study of the common spectrum of a family of commuting dissipative operators with finite-dimensional imaginary component.

An initial attempt to apply the results of the theory of functions on Riemann surfaces to the study of nonselfadjoint operators by means of dilations was undertaken in the work of Abrahamse and Douglas [11].

The present manuscript contains four chapters and an appendix.

The first chapter is concerned with strongly continuous isometric representations of n-parameter semigroups $K \subset R^n$ in Hilbert space. A description is obtained of all representations up to unitary equivalence(§1). A relation is found between these representations and semigroups of shift operators in spaces of solutions of hyperbolic partial differential equations with constant coefficients (§2). The second chapter examines the simplest examples of isometric representations, leading to problems in the theory of functions on a Riemann surface other than the sphere. The third chapter considers multi-parameter semigroups of contractions which admit dilations. Core unitary dilations are introduced. The uniqueness of such dilations within unitary equivalence is proved. In the first section of the fourth chapter, the existence of core unitary dilations for semigroups of the form $\exp(iA_1 t_1 + iA_2 t_2)$ where A_1 and A_2 are bounded commuting operators is proved. A functional model is obtained, analogous to the Nagy-Foias model. The second section studies multi-parameter semigroups of contractions with finite rank of nonunitarity. It is shown that their restrictions to one of their common invariant subspaces of finite codimension form a semigroup of contractions which admit unitary dilations. Furthermore, it is shown that, by "disregarding finite-dimensional effects," the study of such a semigroup of contractions can be replaced by the study of semigroups of projections of multiplication operators on Riemann surfaces. In particular, the von Neumann inequality holds in a subspace of finite codimension. As shown in Chapter 4, the absence of core unitary dilations for n-parameter semigroups of contractions when $n \geq 3$ is caused by the nonextendability to all R^n of some solutions of overdetermined systems of partial differential equations of hyperbolic type. In the appendix of the paper triangular models of pairs of commuting operators are considered. It is shown that the parameters of triangular models are related to nonlinear differential equations. In a special case which is considered in the second chapter, this equation can be integrated by quadratures.

In this final section is obtained a relation which clarifies many further constructions and results. Recall [4, p. 276] that the characteristic operator-function of a bounded operator A_1 was introduced by M. S. Livšic by the formula

$$S(z) = I + 2i \, \text{sgn}(\text{Im } A_1) \, |\text{Im } A_1|^{\frac{1}{2}} \, (A_1^* - z I)^{-1} \, |\text{Im } A_1|^{\frac{1}{2}}\Big|_G .$$

It is shown that if A_1 and A_2 are bounded commuting operators such that

$$|\text{Im } A_2| \leq \text{const} \times |\text{Im } A_1| ,$$

then

$$(\sigma z + \gamma') \, \text{sgn}(\text{Im } A_1) \, S(z) = S(z) \, (\sigma z + \gamma'') \, \text{sgn}(\text{Im } A_1) \quad (3)$$

where

$$\sigma = |A_1 - A_1^*|_G^{-\frac{1}{2}} \, \frac{A_2 - A_2^*}{i} \, |A_1 - A_1^*|_G^{-\frac{1}{2}} ,$$

$$\gamma' = |A_1 - A_1^*|_G^{-\frac{1}{2}} \, \frac{A_1 A_2^* - A_2 A_1^*}{i} \, |A_1 - A_1^*|_G^{-\frac{1}{2}} ,$$

$$\gamma'' = |A_1 - A_1^*|_G^{-\frac{1}{2}} \, \frac{A_2^* A_1 - A_1^* A_2}{i} \, |A_1 - A_1^*|_G^{-\frac{1}{2}} .$$

For the verification of equation (3), it is sufficient to substitute the series

$$S(z) = I - 2i \, \text{sgn}(\text{Im } A_1) \, |\text{Im } A_1|^{\frac{1}{2}} \sum_{k=0}^{\infty} \frac{A_1^{*k}}{z^{k+1}} \, |\text{Im } A_1|^{\frac{1}{2}}$$

in each side of the equation and compare coefficients of corresponding powers of z.

In the case dim $G < \infty$, the eigensubspaces of the operators $(\sigma z + \gamma') \, \text{sgn}(\text{Im } A_1)$ form a fibering over the algebraic set

$$\{(z_1, z_2): \det((\sigma z_1 + \gamma') \, \text{sgn}(\text{Im } A_1)|_G - z_2) = 0\}$$
$$= \{(z_1, z_2): \det((\sigma z_1 + \gamma'') \, \text{sgn}(\text{Im } A_2)|_G - z_2) = 0\}$$
$$= \{(z_1, z_2): d(z_1, z_2) = 0\}$$

with a finite number of discarded "singular" points. The eigensubspaces of the operators $(\sigma z + \gamma'') \, \text{sgn}(\text{Im } A_1)$ form another such fibering. Because of equation (3), the operator-function S(z) generates a complex-analytic morphism of these Hermitian fiberings.

The description of the complex-analytic morphism of the fiberings exhibits the most complete known multi-parameter analog of the characteristic operator-function of a bounded operator.

Only dissipative operators and contractive semigroups are studied in this paper. The characteristic operator-functions of nondissipative commuting operators are studied in the appendix to [16], written by A. U. Abramovič and B. A. Zolotarew.

CHAPTER 1

MULTI-PARAMETER SEMIGROUPS OF ISOMETRIES IN HILBERT SPACE

§1. Functional Models of Isometric Representations

Let K be an open convex cone in R^n. That means that $a_1 x_1 + a_2 x_2$ belongs to K whenever x_1, x_2 belong to K and a_1, a_2 are positive numbers. Assume that for some positive number N

$$\{(t_1,\ldots,t_n) \in R^n : (\sum_2^n |t_j|^2)^{\frac{1}{2}} < N^{-1} t_1\}$$
$$\subset K \subset \{(t_1,\ldots,t_n) \in R^n : (\sum_2^n |t_j|^2)^{\frac{1}{2}} < N t_1\}.$$

The cone K forms a semigroup under addition.

A representation $x \to T(x)$ of this semigroup into the set of bounded linear operators on a Hilbert space H is said to be weakly continuous if

1) The identity
$$T(x_1 + x_2) = T(x_1) T(x_2)$$
holds for all $x_1, x_2 \in K$.

2) The function $(T(x) h_1, h_2)$ is continuous for all h_1, h_2.

3) The identity
$$\lim_{x \to 0} (T(x) h_1, h_2) = (h_1, h_2)$$
holds for all h_1, h_2.

In what follows all Hilbert spaces are assumed to be separable Hilbert spaces over the complex numbers. The representation $T(x)$ is said to be contractive if $\|T(x)\| \leq 1$. Only such representations are now of interest. For them, as is easily shown, the identity
$$\lim_{x \to 0} T(x) h = \lim_{x \to 0} T(x)^* h = h$$
is satisfied. Conditions 2) and 3) are always assumed to be satisfied in what follows and the words "weakly continuous" will be omitted. The representation $T(x)$ is said to be isometric if $\|T(x) h\| = \|h\|$ and unitary if the operators $T(x)$ are unitary,
$$\|T(x) h\| = \|h\|, \qquad \|T^*(x) h\| = \|h\|.$$

Representations $T_1(x)$, $T_2(x)$ in H_1, H_2 are said to be unitarily equivalent if an invertible isometric operator i exists, i: $H_1 \to H_2$, such that
$$T_2(x) i = i T_1(x).$$

The aim is a detailed description of all isometric representations of the semigroup K within unitary equivalence. An important example of such a representation will first be considered.

Let $\sigma_2, \sigma_3, \ldots, \sigma_n$ be bounded selfadjoint and $\gamma_2, \gamma_3, \ldots, \gamma_n$ be possibly unbounded selfadjoint operators in a Hilbert space E. Assume that the selfadjoint operators $\sigma_j t + \gamma_j$ commute pairwise for all real t, $j = 2, \ldots, n$. That is, their resolvents at nonreal points commute. Then for sufficiently small $\|\sigma_2\|, \ldots, \|\sigma_n\|$, the operators

$$f(z) \to \exp(izt_1 + z \sum_2^n \sigma_j t_j + \sum_2^n \gamma_j t_j) f(z)$$

in the Hardy space $H_+^2(E)$ of holomorphic vector-functions in the upper half-plane with values in E form [1] an isometric representation of the semigroup K. The representation is denoted

$$I(\sigma_2, \ldots, \sigma_n, \gamma_2, \ldots, \gamma_n) .$$

THEOREM 1. Every isometric representation of the semigroup K is unitarily equivalent to the orthogonal sum of a unitary representation and a representation of the form $I(\sigma_2, \ldots, \sigma_n, \gamma_2, \ldots, \gamma_n)$. If U' and U" are unitary representations of the semigroup K and if the representations

$$I(\sigma_2', \ldots, \sigma_n', \gamma_2', \ldots, \gamma_n') \oplus U'$$

and

$$I(\sigma_2'', \ldots, \sigma_n'', \gamma_2'', \ldots, \gamma_n'') \oplus U''$$

are unitarily equivalent, then the identities

$$\sigma_j'' V_1 = V_1 \sigma_j' ,$$
$$\gamma_j'' V_1 = V_1 \gamma_j' ,$$
$$U''(x) V_2 = V_2 U'(x)$$

are satisfied for some invertible isometric operators V_1, V_2.

PROOF OF THEOREM 1. Consider an isometric representation I(x) of the semigroup K. Write $x_0 = (1, 0, \ldots, 0)$. It is not difficult to show that the subspace

$$H_0 = \{h: \lim_{t \to \infty} I(tx_0)^* h = 0\} = \{h: \forall x \in K \lim_{t \to \infty} I(tx)^* h = 0\}$$

and its orthogonal complement are invariant subspaces for the isometries I(x). The restrictions of the operators I(x) to the orthogonal complement of H_0 in H are therefore unitary. The proof of the theorem can therefore be restricted to isometric representations I(x) such that

$$\lim_{t \to \infty} I(tx)^* h = 0$$

for all h in H and all x in K.

Use is made of the known description, within unitary equivalence, of all semigroups of isometries [12]. By this result it can be assumed that H is a subspace of the space $L_+^2(\partial I)$ of square-integrable

vector-functions on the positive half-line with values in some Hilbert space ∂I and that
$$I(tx_0) f(\xi) \to \begin{cases} f(\xi - t) & \xi > t, \ t > 0 \\ 0 & \xi < t, \ t > 0 \end{cases}.$$

The infinitessimal generators of one-parameter semigroups of isometries $I(tx)$ are maximal dissipative operators [1]. Denote them $A(x)$ so that
$$I(tx) = \exp itA(x) .$$
A standard "smoothing" procedure
$$h \to \int_K \varphi(x) \, I(x)^* \, h \, dx$$
shows that the intersection $\mathcal{D}_0$ of the domains of the operators is dense in $L_+^2(\partial I)$ and the restriction of $A(x)^*$ to $\mathcal{D}_0$ is a linear operator-function. That means that the nonnegative sesquilinear Hermitian form defined on $\mathcal{D}_0$ by the formula
$$\mu(x,h_1,h_2) = \frac{1}{i} \{(h_1, A(x)^* h_2) - (A(x)^* h_1, h_2)\}$$
is linear in the variable x. Moreover it is clear that
$$\mu(x,h_1,h_2) = -\frac{d}{dt}(I(tx)^* h_1, I(tx)^* h_2)$$
at the point $t = 0$. In particular,
$$\mu(x_0,h_1,h_2) = (h_1(0),h_2(0))_{\partial I} .$$
The inequality
$$\mu((t_1,\ldots,t_n),h_1,h_2) \le (1 + N^2) \, \mu((t_1,0,\ldots,0),h_1,h_2)$$
is needed. It follows from the linearity of the function $\mu(x,h,h)$ in the variable x, the inequality
$$\mu(x,h,h) \ge 0$$
which is valid for all x in K, and the relation
$$((1 + N^2) \, t_1, 0, \ldots, 0) - (t_1, t_2, \ldots, t_n) \in K$$
which is a consequence of the inequality
$$(\sum_2^n t_j^2)^{\frac{1}{2}} < N^{-1} (N^2 \, t_1) .$$

Thus bounded selfadjoint operators $\sigma_2, \ldots, \sigma_n$ exist in the space ∂I such that the identity
$$\mu((t_1,\ldots,t_n),h_1,h_2) = ((t_1 + \sum_2^n \sigma_j t_j) h_1(0), h_2(0))_{\partial I}$$
holds for all h_1, h_2 in $\mathcal{D}_0$.

With every complex number z in the upper half-plane is associated a subspace L_z of $L_+^2(\partial I)$ of vector-functions of the form
$$h(\xi) = u \, e^{iz\xi}$$
with u in ∂I. It is clear that
$$I(x)^* L_z \subset L_z .$$

Furthermore $\mathcal{B}_0 \cap L_z$ is equal to the intersection of the domains of the infinitessimal generators of the one-parameter semigroups obtained on restricting $I(tx)*$ to L_z. In particular, $\mathcal{B}_0 \cap L_z$ is a dense vector subspace of L_z. A maximal dissipative operator $A(x,z)$ is defined in the space ∂H by the formula

$$I(tx)*_{|L_z} (u\, e^{izt}) = (\exp(-i\, A(x,z)*t)\, u)\, e^{izt}.$$

If $h_1 = u_1 e^{izt}$ and $h_2 = u_2 e^{izt}$, then in view of the identities

$$\mu(x_0, h_1, h_2) = (u_1, u_2)_{\partial I} \|e^{izt}\|^2_{L^2(0,\infty)} = (u_1, u_2)_{\partial I} (2\, \text{Im}\, z)^{-1},$$

one obtains

$$((t_1 + \sum_2^n \sigma_j t_j)\, u,\, u)$$
$$= \frac{1}{z - \bar{z}} \{(u,\, A((t_1,\ldots,t_n),z)*\, u) - (A((t_1,\ldots,t_n),z)*\, u, u)\}$$

when $u\, e^{izt}$ belongs to $\mathcal{B}_0$. That is, on letting $(t_1, \ldots, t_n) = x$, one obtains

$$\text{Im}\, ((t_1 + \sum_2^n \sigma_j t_j) z\, u, u) = \text{Im}\, (-A(x,z)*\, u, u). \tag{4}$$

This identity holds for all vectors in the domain of the operator $A(x,z)*$. Thus the vector subspace L of those $u \in \partial I$ such that $u\, e^{izt} \in \mathcal{B}_0$ is norm-dense in the graph of $A(x,z)*$, and the Hermitian form $\text{Im}\, (A(x,z)*\, u, u)$, restricted to L, is bounded.

Thus the inequality

$$\text{Im}\, (A(x,z)*\, u, u) \leq (1 + N^2)\, \text{Im}\, z\, \|u\|^2_{\partial I}$$

holds when $u\, e^{izt} \in \mathcal{B}_0$.

Now from equality of the real parts of the operator-functions

$$-i\, A(x,z)*, \quad i\, (t_1 + \sum_2^n \sigma_j t_j)\, z,$$

which are holomorphic in the variable z, one would like to conclude that their difference is a selfadjoint operator which does not depend on z. The following result holds.

LEMMA 1. The operator function

$$A(x,z) - (t_1 + \sum_2^n \sigma_j t_j)\, z$$

is independent of z and is equal to $\sum_2^n \gamma_j t_j$ for selfadjoint operators $\gamma_2, \ldots, \gamma_n$ where $x = (t_1, \ldots, t_n)$.

The proof of the lemma is presented in the first supplement to Chapter 1. A proof of the first assertion of the theorem will be given.

Pass from the vector-functions $f(\xi) \in L^2_+(\partial I)$ to their Fourier transforms $f^\sim(z) \in H^2_+$. Then

$$I((t_1, \ldots, t_n))\, f^\sim = \exp\, i(zt_1 + z \sum_2^n \sigma_j t_j + \sum_2^n \gamma_j t_j)\, f^\sim \tag{5}$$

by the construction of the operator-function A(x,z) and by Lemma 1. The first assertion of Theorem 1 now follows from the pairwise commutativity of the operators I(x).

For the second assertion of Theorem 1 it is sufficient, as already mentioned, to verify in the hypothesis that the representations U', U" are absent (are of dimension zero). But, as is well known, every operator which realizes the unitary equivalence of the semigroup of operators of multiplication by e^{izt} in the spaces $H^2(E')$, $H^2(E")$ is of the form $f(z) \to V f(z)$ where V is an isometric operator of E' onto E". This completes the proof of Theorem 1.

The definition of the unitary dilation of an isometric representation of the semigroup K will be given and the question of existence and uniqueness of unitary dilations will be studied.

A unitary representation U(x) of the semigroup K in a space H_0 is said to be a unitary dilation of an isometric representation I(x) of this semigroup in a space H if $H \subset H_0$ and I(x) coincides with the restriction of U(x) to H for all x in K.

The unitary dilation U(x) is said to be minimal if the linear span of the vectors $U(x)^* h$, $h \in H$, is dense in the space H_0. If $U_1(x)$, $U_2(x)$ are minimal unitary dilations of the representation I(x), then the identity

$$P_H U_1(y_1 - y_2)_{|H} = P_H U_2(y_1 - y_2)_{|H}$$

holds for all y_1 and y_2 in K since

$$P_H U_{1,2}(y_1 - y_2)_{|H} = P_H U_{1,2}(-y_2) U_{1,2}(y_1)_{|H}$$
$$= P_H U_{1,2}(y_2)^*_{|H} P_H U_{1,2}(y_1)_{|H} = I(y_2)^* I(y_1) .$$

From this follows [1, Theorem 1.71] the uniqueness of the minimal unitary dilation of the isometric representation, that is the existence of a unitary operator V such that

$$U_1(x) V = V U_2(x) .$$

The existence of a unitary dilation of the semigroups Z_+^n, R_+^n is given by a theorem of Ito [1]. In the case $K \neq R_+^n$ the existence of a dilation can be established for example with the help of Theorem 1 since the operators

$$f(\xi) \to \exp i(\xi t_1 + \xi \sum_2^n \sigma_j t_j + \sum_2^n \gamma_j t_j) f(\xi)$$

in the space $L^2(E)$ form a unitary dilation of the representation $I(\sigma_2, \ldots, \sigma_n, \gamma_2, \ldots, \gamma_n)$.

The remainder of this section is concerned with the common invariant subspaces of operators of the minimal unitary dilation $U(x): H_0 \to H_0$ of an isometric representation $I(x): H \to H$.

It can be assumed that the given isometric representation is obtained from $I(\sigma_2,\ldots,\sigma_n,\gamma_2,\ldots,\gamma_n)$ and a unitary representation V of the semigroup K in a space E_2 and is defined in the space $L^2(E) \oplus E_2$ by the formula

$$f(\xi) \oplus h' \to \exp i(\xi t_1 + \xi \sum_2^n \sigma_j t_j + \sum_2^n \gamma_j t_j) \, f(\xi) \oplus V h'$$

and $H = H^2(E)$. If L is a common invariant subspace of the operators $U(x)$, then Theorem 1 applies to the semigroup of restricted transformations $U(x)_{|L}$.

The representation $U(x)_{|L}$ is unitarily equivalent to a representation $I(\sigma_2',\ldots,\sigma_n',\gamma_2',\ldots,\gamma_n') \oplus V'$ in the space $H^2(E') \oplus E_1'$. Let $U'(x)$ be the unitary dilation of the representation $U(x)_{|L}$ in the space $H_0' = L^2(E') \oplus E_1'$.

Because of the uniqueness of the minimal unitary dilation of an isometric representation, the embedding

$$I(\sigma_2',\ldots,\sigma_n',\gamma_2',\ldots,\gamma_n') \oplus V' \to U(x)$$

extends uniquely to an embedding $\theta: U'(x) \to U(x)$. It is clear that the invariant subspace coincides with the range of the operator θ. The following result has been derived.

COROLLARY 1 TO THEOREM 1. If $I(x)$ is the "model" isometric representation of the semigroup K in the space $H_+^2(E) \oplus E_1$, then the operators of its unitary dilation are specified by the formula

$$U(x): f(\xi) \oplus h' \to \exp i(\xi t_1 + \xi \sum_2^n \sigma_j t_j + \sum_2^n \gamma_j t_j) \, f(\xi) \oplus V h'$$

in the space $L^2(E) \oplus E_1$. The common invariant subspaces of the operators $U(x)$ are of the form $\theta(H^2(E') \oplus E'')$ where θ is an isometric operator "intertwining" with the unitary representation

$$f(\xi) \oplus h' \to \exp i(\xi t_1 + \xi \sum_2^n \sigma_j' t_j + \sum_2^n \gamma_j' t_j) \, f(\xi) \oplus V' h' \, .$$

DEFINITION. A common invariant subspace of the operators $U(x)$ is said to be a simply invariant subspace if it contains no nontrivial reducing subspace and is contained in no nontrivial reducing subspace of $U(x)$. A measurable operator-function $\theta(\xi): E' \to E$, which is defined on the real line, is said to be an inner operator-function if it has unitary values for almost all real ξ. No distinction is ordinarily made between inner operator-functions which differ on the left by a constant unitary factor.

The following result is obtained without difficulty from Corollary 1.

COROLLARY 2 TO THEOREM 1. The formula
$$L = \theta(\xi) \, H^2(E')$$
establishes a one-to-one correspondence between the set of all simply invariant subspaces of the operators
$$U((t_1,\ldots,t_n)): f(\xi) \to \exp i(\xi t_1 + \xi \sum_2^n \sigma_j t_j + \sum_2^n \gamma_j t_j) f(\xi)$$
in the space $L^2(E)$ and the set of all inner operator-functions on the real line such that the identity
$$(\xi \sum_2^n \sigma_j t_j + \sum_2^n \gamma_j t_j) \, \theta(\xi) = \theta(\xi) \, (\xi \sum_2^n \sigma_j' t_j + \sum_2^n \gamma_j' t_j)$$
holds for some bounded selfadjoint operators $\sigma_2', \ldots, \sigma_n'$ and for some selfadjoint operators $\gamma_2', \ldots, \gamma_n'$ in the space E'.

Note that from this equation, from the invertibility of $\theta(\xi)$, and from the equation
$$[\sigma_j \xi + \gamma_j, \, \sigma_k \xi + \gamma_k] = 0$$
for all j and k, it follows that the equation
$$[\sigma_j'\xi + \gamma_j', \, \sigma_k'\xi + \gamma_k'] = 0$$
holds for all j and k.

The representation $U((t_1,\ldots,t_n))_{|L}$ is unitarily equivalent to the representation $I(\sigma_2',\ldots,\sigma_n',\gamma_2',\ldots,\gamma_n')$.

REMARK 1. The space $L^2(E)$ can be replaced by the space $H^2(E)$ in the statement of Corollary 2, but then it is necessary to add the condition of holomorphy and boundedness in the upper half-plane of the operator-function $\theta(z)$ with boundary values $\theta(\xi)$.

Finally a description is given of all common simply invariant subspaces of the operators
$$U(t_1,t_2): f(\xi_1,\xi_2) \to f(\xi_1 - t_1, \xi_2 - t_2)$$
in the Hilbert space H_0 of (equivalence classes of) measurable functions of two variables, satisfying the identity
$$f(\xi_1,\xi_2) = f(\xi_1, \xi_2 + 2\pi)$$
and the inequality
$$\|f\|_{H_0}^2 = \int_0^{2\pi} \int_{-\infty}^{+\infty} |f|^2 \, d\xi_1 d\xi_2 < \infty.$$

The simplest example of such an invariant subspace is the subspace L of functions which vanish almost everywhere outside of the region $\{(\xi_1,\xi_2): \xi_1 > 0\}$. Consider the isometric representation $U(t_1,t_2)_{|L}$. The Fourier transformation in the variable ξ_1 realizes the unitary equivalence of the representation $U(t_1,t_2)_{|L}$ and the "model" representation with the operator σ_2 equal to zero and the operator γ_2 unitarily equivalent to the operator $-i\frac{d}{dx}$ in the space $L^2(R/2\pi Z)$ of square-integrable functions on the circle.

Because of Corollary 2, the simply invariant subspaces are described by inner operator-functions $\theta(\xi)$ in $L^2(R/2\pi Z)$ such that

$$\theta(\xi)^{-1}(-i\frac{d}{dx})\theta(\xi) = \sigma\xi + \gamma .\qquad (6)$$

Let us show that the operator γ is unitarily equivalent to the operator $-i\frac{d}{dx}$ in $L^2(R/2\pi Z)$. It is clear from (6) that the spectrum of the operator γ is contained in $2|\xi|\ell^{-1}$-neighborhood of the spectrum of the operator $-i\frac{d}{dx}$ and that the spectrum of the operator $-i\frac{d}{dx}$ is contained in a $2|\xi|\ell^{-1}$-neighborhood of the spectrum of γ. As ξ goes to zero one finds that the spectrum of the operator γ coincides with the set Z of whole numbers. It remains to establish the simplicity of the spectrum of the operator γ.

If some eigenvalue λ of the operator γ were multiple, then for $|\xi| \leq \frac{1}{2}\|\sigma\|^{-1}$, the inequality

$$\|(\gamma + \sigma\xi - \lambda)h\|^2 < \tfrac{1}{2}\|h\|^2$$

would be satisfied in some two-dimensional subspace. That means that the interval $(\lambda - \frac{1}{2}, \lambda + \frac{1}{2})$ would correspond to a not one-dimensional spectral subspace for the operator $\gamma + \sigma\xi$ which is unitarily equivalent to the operator $-i\frac{d}{dx}$ for almost all real x. This yields the desired contradiction.

This shows that the operator γ is unitarily equivalent to the operator $-i\frac{d}{dx}$.

Since the operator γ is determined within unitary equivalence by its invariant subspaces, it can be assumed that $\gamma = -i\frac{d}{dx}$.

Because of the inequality

$$\|\exp i(t_1 z + \sigma t_2 z + \gamma t_2)\| \leq 1$$

when $|t_2| < \ell\, t_1$, the norm of the operator σ does not exceed ℓ^{-1}.

Thus the following result has been proved.

COROLLARY 3 TO THEOREM 1. Every common simply invariant subspace of the operators $U(t_1, t_2)$ with $|t_2| < \ell\, t_1$ is of the form

$$\mathcal{J}\,\theta(\xi)\,H_+^2(L^2(R/2\pi Z))$$

where

$$\theta(\xi): L^2(R/2\pi Z) \to L^2(R/2\pi Z)$$

is an inner operator-function and the operators

$$\xi^{-1}\theta(\xi)^{-1}(-i\tfrac{d}{dx})\theta(\xi),\qquad \xi^{-1}(-i\tfrac{d}{dx})$$

have the same domain, their difference is independent of ξ, and

$$\|\xi^{-1}\{\theta(\xi)^{-1}(-i\tfrac{d}{dx})\theta(\xi) - (-i\tfrac{d}{dx})\}\| \leq \ell^{-1} .\qquad (7)$$

REMARK 3. The description of the common simply invariant subspaces L of the operators $U(t_1,t_2)$, where $(t_1,t_2) \in R \times R$, can be obtained from the decomposition of the spaces L, H into orthogonal sums of subspaces of functions of the form
$$\exp(int_2) f(t_1)$$
for an integer n on applying the classical Beurling-Lax theorem. According to this the invariant subspaces are in one-to-one correspondence with an inner operator-function θ such that
$$(-i \tfrac{d}{dx}) \theta = \theta (-i \tfrac{d}{dx}) \cdot$$
Note that this last identity is obtained from (7) in the limit $\ell \to \infty$.

§2. Isometric Representations and Hyperbolic Systems of Equations.

An essential weakness of the model of an isometric representation constructed in §1 is the use of a distinguished one-parameter semigroup in K. The aim of §2 is the construction of a model which is free from this weakness.

For simplicity the discussion is restricted at the start to a "two-dimensional semigroup" $K \subset R^2$.

The necessary definitions will be given. Let E be a locally convex topological vector space over the field of complex numbers which is metrizable and complete. A sesquilinear form $\gamma(h_1,h_2)$, defined on a dense vector subspace $\mathcal{B}(\gamma) \subset E$, is said to be Hermitian if $\operatorname{Im} \gamma(h,h) = 0$ and separately continuous if the linear functionals $\gamma(h,\cdot)$ and $\gamma(\cdot,h)$ extend by continuity to the whole space E for every element h of $\mathcal{B}(\gamma)$. A separately continuous Hermitian form γ is said to be selfadjoint if every vector $g \in E$ for which the linear functional $h \to \gamma(h,g)$ extends by continuity to all of E belongs to $\mathcal{B}(\gamma)$.

Assume now that a selfadjoint sesquilinear form γ and continuous selfadjoint forms σ_1, σ_2 are given in the space E. Assume that for all $(a_1,a_2) \in K$,
1) $a_1 \sigma_1 + a_2 \sigma_2 \geq 0$,
2) the sets
$$\{h: a_1 \sigma_1(h,h) + a_2 \sigma_2(h,h) < \epsilon\}$$
form a basis for the neighborhoods of the origin in the space E.

Note that the sesquilinear forms σ_1, σ_2, γ are to be identified with the conjugate linear transformations $h \to \sigma_1(\cdot,h)$, $h \to \sigma_2(\cdot,h)$, $h \to \gamma(\cdot,h)$ of the space E into the dual space E'.

The model isometric representation of the semigroup K will be constructed in the space of solutions of the hyperbolic system of

equations
$$\{\sigma_2(-i\frac{\partial}{\partial x_1}) - \sigma_1(-i\frac{\partial}{\partial x_2}) + \gamma\} u = 0 \ . \tag{8}$$

The following properties of a solution $u(\xi_1,\xi_2) \in \mathcal{C}_0^\infty(\mathcal{B}(\gamma))$ of equation (8) are easily verified:

1) If $u(t,0) = 0$ for $t \le \epsilon$, then
$$u(t\xi_1, t\xi_2) = 0$$
for $t \le \text{const}(\epsilon, \xi_1, \xi_2)$.

2) The integral
$$\int \sigma_1(u,u) \, d\xi_1 + \sigma_2(u,u) \, d\xi_2$$
taken over the line $(t\xi_1, t\xi_2)$ is independent of the choice of $(\xi_1,\xi_2) \in K$.

Assertion 2) allows one, on completing the vector space of solutions in $\mathcal{C}_0^\infty(\mathcal{B}(\gamma))$ in the energy norm
$$\sqrt{[u,u]} \ ,$$
to obtain a unitary representation
$$f(\xi_1,\xi_2) \to f(\xi_1 - t_1, \xi_2 - t_2)$$
of the group R^2 in the Hilbert space $H_{\sigma_1,\sigma_2,\gamma}$ of solutions of the hyperbolic system (8). Assertion 1) shows that the completion of the solutions $f \in \mathcal{C}_0^\infty(\mathcal{B}(\gamma))$ which vanish for $\xi_2 = 0$, $\xi_1 < \epsilon_f$, $\epsilon_f > 0$ in the energy norm is a common invariant subspace $H_{+\sigma_1,\sigma_2,\gamma}$ of the operators
$$f(\xi_1,\xi_2) \to f(\xi_1 - t_1, \xi_2 - t_2)$$
with (t_1,t_2) in K. The restrictions of these operators to $H_{+\sigma_1,\sigma_2,\gamma}$ form an isometric representation of the semigroup K, denoted in what follows $J_{\sigma_1,\sigma_2,\gamma}(t_1,t_2)$.

THEOREM 2. Every isometric representation of the semigroup K is unitarily equivalent to the orthogonal sum of a unitary representation and a representation of the form $J_{\sigma_1,\sigma_2,\gamma}(t_1,t_2)$. The representations
$$J_{\sigma_1',\sigma_2',\gamma'}(t_1,t_2) \oplus U' \ , \quad J_{\sigma_1'',\sigma_2'',\gamma''}(t_1,t_2) \oplus U''$$
are unitarily equivalent if, and only if, the unitary representations U', U" are unitarily equivalent and a linear homeomorphism $\varphi: E' \to E''$ exists for which
$$\sigma_1''(\varphi h, \varphi h) = \sigma_1'(h,h) \ , \quad \sigma_2''(\varphi h, \varphi h) = \sigma_2'(h,h) \ ,$$
$$\varphi \mathcal{B}(\gamma') = \mathcal{B}(\gamma'') \ , \quad \gamma''(\varphi h, \varphi h) = \gamma'(h,h) \ .$$

The proof of Theorem 2 is given in the second supplement to Chapter 1.

Now for the construction, for a given isometric representation $I(t_1,t_2)$ of the semigroup K in the space H, of the quantities E, σ_1, σ_2, γ appearing in the decomposition

$$I(t_1,t_2) \approx J_{\sigma_1,\sigma_2,\gamma}(t_1,t_2) \oplus U(t_1,t_2) \;.$$

Consider maximal dissipative operators $A(a_1,a_2)$, $(a_1,a_2) \in K$, such that

$$I(a_1 t, a_2 t)^* = \exp\{-it\, A(a_1,a_2)^*\} \oplus U(a_1 t, a_2 t)^* \;.$$

Use is made of the fact that any strongly continuous one-parameter semigroup of contractions is of the form $\exp(iAt)$ where A is a maximal dissipative operator. It is not difficult to show that on the intersection $\mathcal{D}$ of the domains of the operators $A(a_1,a_2)^*$, which is dense in the space H, nonnegative Hermitian forms are defined by the identity

$$(a_1 \sigma_1 + a_2 \sigma_2)(h,h) = 2 a_1 \operatorname{Im}(h, A_1^* h) + 2 a_2 \operatorname{Im}(h, A_2^* h) \;.$$

As in the proof of Theorem 1, one obtains

$$(a_1 \sigma_1 + a_2 \sigma_2)(h,h) \le \text{const} \times a_1\, \sigma_1(h,h) \;.$$

It allows a factorization of $\mathcal{D}_0$ with respect to the kernel of the hermitian form $a_1 \sigma_1 + a_2 \sigma_2$ in order to complete the factor space with respect to the norm

$$\sqrt{a_1 \sigma_1(h,h) + a_2 \sigma_2(h,h)} \;,$$

obtaining thereby a topological vector space E which does not depend on the choice of element $(a_1,a_2) \in K$. The Hermitian forms σ_1, σ_2 generate Hermitian forms on E. The natural mapping $\mathcal{D}_0 \to E$ will be denoted φ. In the particular case of the model representation

$$(J_{\sigma_1',\sigma_2',\gamma'} \oplus U')(t_1,t_2) \;,$$

the operator φ can be exhibited as $u(x_1,x_2) \to u(0,0)$ with $E = E'$, $\sigma_{1,2} = \sigma_{1,2}'$. Then E, σ_1, σ_2 are constructed. Thus

$$(a_1 \sigma_1 + a_2 \sigma_2)(\varphi h, \varphi h) = 2 a_1 \operatorname{Im}(h, A_1^* h) + 2 a_2 \operatorname{Im}(h, A_2^* h) \;. \tag{9}$$

Let now $h \in \mathcal{D}_0$, $g \in \mathcal{D}_0$, and

$$u(\xi_1,\xi_2) = \varphi\, (J_{\sigma_1',\sigma_2',\gamma'} \oplus U')^* h$$

Then

$$\sigma_2(\tfrac{\partial u}{\partial \xi_1}, \varphi\, g) - \sigma_1(\tfrac{\partial u}{\partial \xi_2}, \varphi\, g)$$

$$= \sigma_2(\varphi\, (-iA_1^*)\, h, \varphi\, g) - \sigma_1(\varphi\, (-iA_2^*)\, h, \varphi\, g)$$

$$= \frac{1}{i}\, ((-iA_1^* h, A_2^* g) - (-iA_2^* h, A_1^* g))$$

In this way is obtained the identity

$$\gamma(\varphi\, h_1, \varphi\, h_2) = \frac{1}{i}\, \{(A_2^* h_1, A_1^* h_2) - (A_1^* h_1, A_2^* h_2)\} \;. \tag{10}$$

Formula (10) defines a Hermitian form γ on $\varphi \mathcal{B}_0$. It remains to note that the range of the operator φ coincides with the domain of the form γ in the case of a representation of the form $I_{\sigma_1,\sigma_2,\gamma} \oplus U$ and therefore also in the general case.

An isometric representation of a semigroup $K \subset R^n$ with $n > 2$ can also be obtained in the space of solutions of a hyperbolic system of equations with constant coefficients. The only difference from a two-parameter semigroup, which has been studied in detail, consists of the overdetermination of the system of equations then arising. As in the case $n = 2$, the space of solutions turns out to be isomorphic to the space of restrictions to the line $(t\xi_1, t\xi_2, \ldots, t\xi_n)$ where $(\xi_1, \xi_2, \ldots, \xi_n) \in K$. But in the case $n > 2$, the solubility of the Cauchy problem with "arbitrary" initial data on such a line leads to additional conditions on the coefficients of the system of equations

$$\sigma_j (-i \frac{\partial u}{\partial \xi_k}) - \sigma_k (-i \frac{\partial u}{\partial \xi_j}) + \gamma_{jk} = 0,$$

where σ_k, γ_{jk} are selfadjoint sesquilinear forms and $\gamma_{jk} = -\gamma_{kj}$. The compatibility conditions are written as continuity hypotheses on the forms γ_{jk}. They are equivalent to the commutativity of the operators

$$u \to \sigma_k^{-1} \sigma_j (-i \frac{\partial u}{\partial \xi_k}) + \sigma_k^{-1} \gamma_{jk} u,$$

$$u \to \sigma_k^{-1} \sigma_\ell (-i \frac{\partial u}{\partial \xi_k}) + \sigma_k^{-1} \gamma_{\ell k} u$$

and are of the form

1) $\sigma_j \sigma_k^{-1} \sigma_\ell = \sigma_\ell \sigma_k^{-1} \sigma_j$
2) $\sigma_j \sigma_k^{-1} \gamma_{\ell k} - \gamma_{jk} \sigma_k^{-1} \sigma_\ell = \sigma_\ell \sigma_k^{-1} \gamma_{jk} - \gamma_{\ell k} \sigma_k^{-1} \sigma_j$
3) $\gamma_{jk} \sigma_k^{-1} \gamma_{\ell k} = \gamma_{\ell k} \sigma_k^{-1} \gamma_{jk}$.

An important definition will now be given.

The topological vector space obtained on factoring the vector space $\mathcal{B}_0$ with respect to the subspace

$$\{h: \forall x \in K \quad \frac{d}{dt} \left(\|I(tx)* h\|^2 \right)\Big|_{t=1} = 0\}$$

and completing with respect to the norm

$$\sqrt{- \frac{d}{dt} \left(\|I(tx)* h\|^2 \right)\Big|_{t=1}}$$

is called the boundary space ∂I of the isometric representation $I(x)$ of the semigroup K in the space H.

An example is presented which explains the term "boundary space." Assume that $(r_1(t), r_2(t))$ is a smooth curve in the plane such that

$$\frac{dr_1}{dt} > 0, \quad \frac{dr_2}{dt} < 0, \quad -\infty < t < \infty \ .$$

Assume that the curve $r(t)$ divides the plane into two connected components. Denote by Ω the component for which $\Omega + (t_1, t_2) \Omega \subset \Omega$ for $t_1 > 0$, $t_2 > 0$, and by $I_\Omega(t_1, t_2)$ the semigroup of isometries

$$f(\xi_1, \xi_2) \to f(\xi_1 - t_1, \xi_2 - t_2)$$

in the space $L^2(\Omega, d\xi_1 d\xi_2)$. The boundary space is identified in a natural way with the space of equivalence classes of measurable functions on the curve

$$\partial\Omega = \{(r_1(t), r_2(t)) : -\infty < t < \infty\}$$

for which

$$\sigma_2(f,f) = \int_{\partial\Omega} |f|^2 \, d\xi_1 < \infty \ ,$$

$$\sigma_1(f,f) = \int_{\partial\Omega} |f|^2 \, d\xi_2 < \infty \ .$$

In the transition to the model representation, the function $f(\xi_1, \xi_2) \in L^2(\partial\Omega, d\xi_1 d\xi_2)$ corresponds to the function $F(t, t_1, t_2)$ of three variables defined by the formula

$$F = f(r_1(t) + t_1, r_2(t) + t_2) \ .$$

The partial derivatives of the function $F(t, t_1, t_2)$ are equal to

$$\frac{\partial F}{\partial t} = \frac{\partial f}{\partial \xi_1} \frac{dr_1}{dt} + \frac{\partial f}{\partial \xi_2} \frac{dr_2}{dt}, \quad \frac{\partial F}{\partial t_1} = \frac{\partial f}{\partial \xi_1}, \quad \frac{\partial F}{\partial t_2} = \frac{\partial f}{\partial \xi_2} \ .$$

The hyperbolic system of equations is of the form

$$i \frac{dr_1}{dt} \frac{\partial F}{\partial t_1} + i \frac{dr_2}{dt} \frac{\partial F}{\partial t_2} - i \frac{\partial F}{\partial t} = 0$$

and consequently the sesquilinear form γ is equal to

$$\gamma(f,f) = \frac{1}{i} \int_\Omega \overline{f} \, df \ .$$

§3. **Isometric Representations of Finite Rank of Nonunitarity.**

An isometric representation $I(x)$ of the semigroup K in a Hilbert space H is said to be completely nonunitary if $\forall h \in H$

$$\lim_{t \to \infty} \|I(tx)^* h\| = 0 \ .$$

The multiplicity of the spectrum of the minimal unitary dilation of the one-parameter semigroup of isometries $I(tx)$, $x \in K$, is called the rank of nonunitarity of the completely nonunitary isometric representation $I(x)$. The rank of nonunitarity is equal to the dimension of the boundary space ∂I.

§3 will be concerned with isometric representations for which $\dim(\partial I) < \infty$.

In §2 Hermitian-symmetric sesquilinear forms σ_j, γ_{ij} were constructed in the boundary space ∂I. Choose a basis for ∂I and consider the determinant

$$\det(z_i \sigma_j - z_j \sigma_i + \gamma_{ij}).$$

On passage from one basis in the boundary space to another, this function of the variables z_i, z_j is multiplied by a nonzero constant. It follows that the set of zeros of the determinant is independent of the choice of basis in ∂I.

The set

$$\text{spset } I(x) = \{(z_1,\ldots,z_n) \in C^n \colon \text{Im } \Sigma\, a_j z_j > 0,$$
$$(a_1,\ldots,a_n) \in K,\ \det(z_k \sigma_\ell - z_\ell \sigma_k + \gamma_{k\ell}) = 0,\ k,\ell = 1,\ldots,n\}$$

is called the canonical spectral set of the completely nonunitary isometric representation

$$I((t_1,\ldots,t_n)) = \exp i \Sigma\, A_j t_j.$$

This definition is suggested by the inequality

$$\|\psi(A_1,\ldots,A_n)\| \leq \sup_{(\lambda_1,\ldots,\lambda_n) \in \text{spset } I(x)} |\psi(\lambda_1,\ldots,\lambda_n)|$$

which holds for an arbitrary function $\psi(\lambda_1,\ldots,\lambda_n)$ of the form

$$\psi(\lambda_1,\ldots,\lambda_n) = \Sigma\, c_j \exp i \sum_{k=1}^{n} a_{kj} \lambda_k$$

where $(a_{1j},\ldots,a_{nj}) \in K$. The inequality is a consequence of the evident (in view of Theorem 1) inclusion of the spectrum of the minimal unitary dilation of the representation $I(x)$ in the closure of spset $I(x)$.

As is easily shown, the boundary Γ of the canonical spectral set is a real algebraic curve

$$\{(\xi_1,\ldots,\xi_n) \in R^n \colon \det(\xi_i \sigma_j - \xi_j \sigma_i + \gamma_{ij}) = 0,\ i,j = 1,\ldots,n\}$$

which may be reducible. The curve was introduced by M. S. Livšic in his recent work on the theory of operator colligations related to families of commuting nonselfadjoint operators.

An isometric representation is said to be primary if the algebraic curve introduced is irreducible.

Let $\Gamma = \cup\, \Gamma_i$ be the decomposition of the curve Γ into irreducible components and let A_j be the infinitesimal operators such that

$$I(t_1,\ldots,t_n) = \exp i \Sigma\, A_j t_j.$$

Define the primary (root) subspaces H_i by means of polynomials which vanish on the sets $\Gamma \setminus \Gamma_i$

$$H_i = (\bigcap_{\{P\colon P(z_1,\ldots,z_n)|\Gamma\setminus\Gamma_i = 0\}} \text{Ker } P(-A_1^*,\ldots,-A_n^*))^\perp.$$

It is clear that

$$I(t_1,\ldots,t_n) H_i \subset H_i.$$

DEFINITION. The subspaces H_i are called the primary subspaces and the restrictions $I(t_1,\ldots,t_n)_{|H_i}$ are called the primary components of the representation $I(t_1,\ldots,t_n)$.

PROPOSITION 1. The primary subspaces of a completely nonunitary isometric representation of finite rank of nonunitarity are pairwise orthogonal and their sum has finite codimension.

PROOF. If is not difficult to verify the pairwise orthogonality of the primary spaces using the functional model of an isometric representation obtained in §1.

Argue by contradiction, assuming that the dimension of $(\oplus_i H_i)^\perp$ is infinite. Then an infinite-dimensional vector space of smooth bounded vector functions exists on the half-line $x_1 \geq 0$, $x_2 = x_3 = \cdots = x_n = 0$, which are extendable to solutions in R^n of the system of differential equations corresponding to $(\oplus_i H_i)^\perp$. The space of all such solutions is finite-dimensional since the intersection of the algebraic curves has measure zero. A contradiction of the hypothesis that $\dim(\oplus_i H_i)^\perp = \infty$ is obtained. This completes the proof of Proposition 1.

Assume that
$$I(t_1,\ldots,t_n) = \exp i \Sigma A_j t_j$$
is a completely nonunitary primary representation of the semigroup K in a space H and that
$$U(t_1,\ldots,t_n) = \exp i \Sigma \hat{A}_j t_j$$
be its minimal unitary dilation in $H_0 \supset H$.

DEFINITION. By the regularization of a primary representation $I(t_1,t_2,\ldots,t_n)$ is meant its subrepresentation in the space
$$\{h \in H: \forall \varphi \in H^\infty(\text{spset}(I(x))) \quad \varphi(\hat{A}_1,\ldots,\hat{A}_n) h \in H\} .$$
Here $H^\infty(\text{spset } I(x))$ denotes the Banach algebra of holomorphic functions on the set spset $I(x)$ equipped in the natural way with the structure of a Riemann surface. A primary representation is said to be regular if it coincides with its regularization.

Thus in the case of a regular representation the formula
$$\varphi(A_1,\ldots,A_n) = \varphi(\hat{A}_1,\ldots,\hat{A}_n)_{|H}$$
specifies a sufficiently rich functional calculus, $H^\infty(\text{spset } I(x))$.

PROPOSITION 2. The regularization of a primary representation is a representation of finite codimension.

The proof makes use of the functional model of the isometric representation constructed in Section 1. Use is also made of the fact that the spectral projections onto the common eigensubspaces of the operators
$$zt_1 + z \sum_2^n \sigma_j t_j + \sum_2^n \gamma_j t_j$$
define a meromorphic operator-function with a finite number of poles on the Riemann surface spset $I(t_1,\ldots,t_n)$. See [13]. Details of the proof will not be given.

According to Theorem 1, a regular representation is unitarily equivalent to a representation
$$f(z) \to \exp i(zt_1 + z \sum_2^n \sigma_j t_j + \sum_2^n \gamma_j t_j) f(z)$$
in the space $H_+^2(E)$ where dim $E < \infty$. It is not difficult to show that, for all except a finite number of real numbers ξ, the common eigensubspaces of the operators $\sigma_j \xi + \gamma_j$, $j = 2,\ldots,n$ have equal dimension r.

The spectral measure of the family of commuting selfadjoint operators
$$f(\xi) \to (\sigma_j \xi + \gamma_j) f(\xi)$$
is supported on the previously introduced algebraic curve and is absolutely continuous on this curve. The multiplicity of the common spectrum of the infinitessimal operators is uniformly equal to r.

That means that the group of operators
$$\exp i(\xi t_1 + \xi \sum_2^n \sigma_j t_j + \sum_2^n \gamma_j t_j)$$
is unitarily equivalent to the group of operators of multiplication by
$$\exp i \sum_2^n \xi_j t_j$$
in the space
$$L^2(\partial \text{ spset } I(x), d\xi_1) \oplus \cdots \oplus L^2(\partial \text{ spset } I(x), d\xi_1)$$
where the direct sum is taken r times.

The regularity of the representation $I(x)$ means that under this unitary equivalence the space $H_+^2(E)$ corresponds to a common invariant subspace of the operators of multiplication by
$$\varphi(\xi_1,\ldots,\xi_n) \in H^\infty(\text{spset } I(x)) .$$
Recall that, for every point $p \in \text{spset } I(x)$, a measure m_p exists representing it in the space ∂ spset $I(x)$. That is the identity
$$f(p) = \int f \, dm_p$$
holds for all functions $f \in H^\infty(\text{spset } I(x))$ which have continuous extensions to the closure of the set spset $I(x)$. It is not difficult

to show that the measures $(1 + \xi_1^2)^{-1} d\xi_1$ and the representing measures dm_p are related by the inequalities

$$(1 + \xi_1^2)^{-1} d\xi_1 \leq \text{const}(p) \, dm_p,$$
$$dm_p \leq \text{const}(p) \, (1 + \xi_1^2)^{-1} d\xi_1.$$

Therefore the well-known results of Hasumi [5], Voichick [6], and Forelli [11] are applicable to the description of H^∞-invariant subspaces. Let τ be a unitary r-dimensional representation of the fundamental group $\pi_1(\text{spset } I(x))$ of the Riemann surface spset $I(x)$ and let $H^2(dm)$ be the Hardy space on spset $I(x)$. Consider the spaces $H_\tau^2(d\xi_1)$ of holomorphic (multi-valued) vector-functions which, on passage around a contour $r(t) \in \text{spset } I(x)$, are multiplied by $\tau([r])$ and for which

$$(i + \xi_1) \, f(\xi_1) \in H^2(dm).$$

The norm of the Hilbert space H_τ^2 is defined by the formula

$$\|f\|_{H_\tau^2}^2 = \int_{\partial \text{ spset } I(x)} \|f\|_E^2 \, d\xi_1.$$

According to the cited results [5,6,11], invariant subspaces are of the form ψH_τ^2 where ψ is a holomorphic operator-function in spset $I(x)$ with isometric boundary values at almost all point of ∂ spset $I(x)$. The following result has been arrived at.

THEOREM 3. A completely nonunitary regular isometric representation of the semigroup K is unitarily equivalent to a representation by multiplication operators

$$f \to \exp i \left(\sum_{1}^{n} \xi_j t_j \right) f$$

in the space $H_\tau^2(d\xi_1)$ where τ is a unitary representation of the fundamental group of the Riemann surface spset $I(x)$. Two completely nonunitary regular isometric representations are unitarily equivalent if, and only if, their canonical spectral sets coincide and the corresponding representations of the fundamental group are unitarily equivalent.

REMARK. Assume that $f_1(z), f_2(z), \ldots, f_n(z)$ are rational functions which are real for real z such that

$$|\text{Im } f_j| \leq \text{const} \times \text{Im } f_1.$$

Then the operators of multiplication by the functions $\exp i \sum t_j f_j$ form an isometric representation of the semigroup

$$\{(t_1, \ldots, t_n) : |t_j| \leq \text{const} \times t_1, \, j = 2, 3, \ldots, n\}$$

in the Hardy space H_+^2 of holomorphic functions in the upper half-plane. It is not difficult to find such $f_1, \ldots, f_n$ such that the set of common

invariant subspaces of the operators of multiplication by
$$\exp i \Sigma\, t_j f_j$$
does not coincide with the set of invariant subspaces of the operators of multiplication by
$$\exp itz, \quad t \geq 0 \; .$$
Let L be one of the "nonstandard" common invariant subspaces of the operators
$$\exp i \Sigma\, t_j f_j \; .$$
Then their restrictions to L produce an example of an isometric representation which is not regular.

The following result is derived from Theorem 3 using Propositions 1 and 2.

PROPOSITION 3. Let $I(t_1,\ldots,t_n)$ be a completely nonunitary isometric representation of the semigroup K with finite rank of nonunitarity and let $I_\alpha(t_1,\ldots,t_n)\colon H_\alpha \to H_\alpha$, $\alpha = 1,2,\ldots,r$ be the regularizations of the primary components of the representation $I(t_1,\ldots,t_n)$. Then

1) $H_{\alpha'} \perp H_{\alpha''}$ when $\alpha' \neq \alpha''$,

2) the dimension of the orthogonal complement of
$$H_1 \oplus H_2 \oplus \cdots \oplus H_r$$
in H is finite,

3) every representation $I_\alpha(t_1,\ldots,t_n)$ is unitarily equivalent to a representation by multiplication operators
$$f \to \exp i(\sum_1^n \xi_j t_j)\, f$$
in the previously defined space $H^2_{\tau_i}(d\xi_1)$. Here τ_j is a unitary representation of the fundamental group of the Riemann surface
$$\text{spset } I_\alpha(t_1,\ldots,t_n) \; .$$

SUPPLEMENT 1, PROOF OF LEMMA 1. Lemma 1 was stated in the proof of Theorem 1. Consider the operators
$$\gamma(x,z) = A(x,z)^* - (t_1 + \sum_2^n t_j \sigma_j)\, z \; . \tag{11}$$
By the proof of Theorem 1, the operators $\gamma(x,z)$ are symmetric:
$$\gamma(x,z) \subset \gamma(x,z)^* \; .$$
The graph of the operator $\gamma(x,z)$ is closed because the graph of the operator $A(x,z)$ is closed and because the operators σ_j are bounded. Selfadjointness of the operator $\gamma(x,z)$ now follows from the boundedness of the set
$$\{\operatorname{Im} \lambda \colon \lambda \in \operatorname{spec} \gamma(x,z)\}$$

of imaginary parts of points in the spectrum of the operator γ. This last result follows from the boundedness of the set
$$\{\text{Im } \lambda: \lambda \in \text{spec } A(x,z)\}$$
and from the boundedness of the operators σ_j. Finally the boundedness of the set
$$\{\text{Im } \lambda: \lambda \in \text{spec } A(x,z)\}$$
is a consequence of the dissipative property of the operator A and the inequality
$$\exp i((N^2+1)t_1 i \text{ Im } z) \|h\|^2 \leq \|\exp iA(x,z) h\|^2 \tag{12}$$
by means of which it follows that
$$\text{Im } \lambda \leq (N^2+1) t_1 \text{ Im } z \ .$$
Thus
$$\gamma(x,z) = \gamma(x,z)^* \ .$$
In view of (12), the operator-function
$$\int_0^\infty \exp(-itA(x,z)) e^{-t} dt \ ,$$
and therefore the operator-function
$$(A(x,z) - i)^{-1} \ ,$$
is holomorphic for
$$0 < \text{Im } z < (N^2+1)^{-1} t_1^{-1} \ .$$
The operator function
$$(A(x,z) + i)^{-1}$$
is holomorphic in the upper half-plane by the dissipative property of the operators $A(x,z)$. Therefore the identities
$$(A(x,z) + i)^{-1} - (\gamma(x,z) + i)^{-1}$$
$$= (\gamma(x,z) + i)^{-1}(t_1 + \sum_2^n t_j\sigma_j) z (A(x,z) + i)^{-1} \ ,$$
$$(A(x,z) - i)^{-1} - (\gamma(x,z) - i)^{-1}$$
$$= (\gamma(x,z) - i)^{-1}(t_1 + \sum_2^n t_j\sigma_j) z (A(x,z) - i)^{-1}$$
imply that the operator-functions $\gamma(x,z) \pm i$ are holomorphic in a region
$$0 < \text{Im } z < (N^2+1)^{-1}t_1^{-1}; \quad |z| < \epsilon_0, \quad \epsilon_0 > 0 \ . \tag{13}$$
But then
$$\frac{i}{2}\{(\gamma(x,z) + i)^{-1} - (\gamma(x,z) - i)^{-1}\}$$
$$= (\gamma(x,z) - i)^{-1}(\gamma(x,z) + i)^{-1}$$
is a holomorphic function whose values are bounded selfadjoint operators. Thus in the region (13)
$$\gamma(x,z) = \gamma(x)$$
and
$$A(x,z) = (t_1 + \sum_2^n \sigma_j t_j) z + \gamma(x) \ .$$

Therefore the holomorphic operator-functions
$$\exp itA(x,z), \quad \exp it(zt_1 + z\sum_2^n \sigma_j t_j + \gamma(x)),$$
$x = (t_1,\ldots,t_n)$ coincide in the region (13). That means that they are everywhere equal in the upper half-plane. It remains to observe that, for every point z in the upper half-plane, the operators $\exp iA(x,z)$ form a strongly continuous representation of the semigroup K. Therefore the operator-functions $A(x,z)$, and hence also $\gamma(x)$, are linear in x:
$$\gamma(x) = \sum t_j \gamma_j$$
where $x = (t_1,\ldots,t_n)$. But
$$A((t_1,0,\ldots,0),z) = t_1 z .$$
That is, $\gamma_1 = 0$,
$$\gamma = \sum_2^n t_j \gamma_j ,$$
$$A(x,z) = zt_1 + z\sum_2^n t_j \sigma_j + \sum_2^n t_j \gamma_j ,$$
$x = (t_1,\ldots,t_n)$. The selfadjointness of γ_j follows from the previously established selfadjointness of the operator-function
$$\gamma(x,z) = \sum_2^n t_j \gamma_j .$$
The commutativity of the operators
$$zt_1 + z\sum_2^n t_j \sigma_j + \sum_2^n t_j \gamma_j$$
is a consequence of the commutativity of the operators of multiplication by
$$\exp i(zt_1 + z\sum_2^n t_j \sigma_j + \sum_2^n t_j \gamma_j) .$$
This completes the proof of the lemma.

SUPPLEMENT 2. PROOF OF THEOREM 2. By Theorem 1, it is sufficient, for the proof of the first assertion of Theorem 2, to associate with every pair of selfadjoint operators σ, γ for which $\|\sigma\| < \infty$, a representation of the form $J_{\sigma_1',\sigma_2',\gamma'}$ which is unitarily equivalent to a representation
$$f(z) \to \exp i(zt_1 + z\sigma t_2 + \gamma t_2) f(z) .$$
Operators σ_1', σ_2', γ' are introduced by means of the formulas
$$\sigma_1'(h,h) = (h,h), \quad \gamma'(h,h) = (\gamma h, h), \quad \sigma_2'(h,h) = (\sigma h, h) .$$
The hyperbolic system of equations (8) takes the form
$$\{\sigma_2 (-i\frac{\partial}{\partial x_1}) - (-i\frac{\partial}{\partial x_2}) + \gamma\} u = 0 . \tag{14}$$
The space $H_{+\sigma_1',\sigma_2',\gamma'}$ is obtained as the completion of the vector space of smooth finite solutions $f(x_1,x_2)$ of the system (14) for which $f(x_1,0) = 0$ when $x_1 < \epsilon_f$, $\epsilon_f > 0$. An element of the space $H_+^2(E)$ is

associated with every solution f by means of the Fourier transformation. This representation extends by continuity to an isometry

$$i: H^2_{+\sigma_1', \sigma_2', \gamma'} \to H^2_+(E) .$$

The identity

$$i \, J_{\sigma_1', \sigma_2', \gamma'}(t_1, t_2) = \exp i(zt_1 + z\sigma t_2 + \gamma t_2) \, i$$

is not difficult to obtain in the chosen dense vector subspace and it then follows in the whole space. Thus the first assertion of Theorem 2 will be verified if it is possible to show that the transformation i is surjective. In other words, it is sufficient to show that, for every smooth finite vector-function $f_0(x)$ on the line with values in $\mathcal{D}(\gamma) \subset E$ and for every $\epsilon > 0$, a smooth finite solution $F(\xi_1, \xi_2) \in \mathcal{C}_0^\infty(\mathcal{D}(\gamma))$ of the hyperbolic system (14) exists such that

$$\int_{-\infty}^{+\infty} \|F(\xi_1, 0) - f_0(\xi_1)\|_E^2 \, d\xi_1 \leq \epsilon$$

and such that the function $F(\xi_1, 0)$ vanishes outside of an ϵ-neighborhood of the set on which $f_0(\xi_1)$ is nonzero. Indeed the restriction to the set where $F(\xi_1, 0)$ is nonzero allows one to obtain the inclusion

$$\text{supp } F(\xi_1, 0) \subset (0, \infty)$$

if

$$\text{supp } f_0 \subset (0, \infty)$$

and the "Cauchy data" of the solutions $F(\xi_1, \xi_2)$ are dense in $L^2((0, \infty), E)$.

Thus it remains to construct a solution $F(\xi_1, \xi_2)$ corresponding to a given vector-function $f_0(x)$ and a positive number ϵ.

The selfadjointness of the operator

$$H_0 = -i \, \sigma_2 \frac{\partial}{\partial \xi_1} + \gamma$$

in the space $L^2((-\infty, +\infty), E)$ permits the consideration of a "generalized solution" of the Cauchy problem

$$f(\xi_1, \xi_2) = \exp(\xi_2 \sigma_2 \frac{\partial}{\partial \xi_1} + i\xi_2 \gamma) \, f(\xi_1) .$$

Let $\psi_1(t)$, $\psi_2(t)$ be smooth functions, with nonnegative values and total integral one, which vanish outside of a sufficiently small neighborhood of the origin. It will be shown that the function

$$F(\xi_1, \xi_2) = \iint f(\xi_1 - t_1, \xi_2 - t_2) \, \psi_2(t_2) \, \psi_1(t_1) \, dt_2 dt_1$$

is the desired smooth solution of the Cauchy problem.

STEP 1. The domain $\mathcal{D}(H_0)$ of the operator H_0 consists of those vectors h for which

$$\int (1 + t^2) \, d(E_t h, h) < \infty,$$

where E_t is the resolution of the identity. Equip $\mathcal{D}(H_0)$ with the norm

of the graph,
$$\|h\|^2_{\Gamma(H_0)} = \|h\|^2 + \|H_0 h\|^2 = \int (1 + t^2)\, d(E_t h, h) \ .$$
It is clear that
$$e^{iH_0 t} \mathcal{D}(H_0) = \mathcal{D}(H_0) \ .$$
Using the formula
$$((e^{-iH_0 \epsilon} - 1)(1 + H_0^2)(e^{iH_0 \epsilon} - 1)\, h, h)$$
$$= \int (1 + t^2)\, |e^{i\epsilon t} - 1|^2\, d(E_t h, h)$$
and applying the Lebesgue dominated convergence theorem in the limit $\epsilon \to 0$, one obtains without difficulty the strong continuity of the group of unitary operators
$$e^{iH_0 t}|_{\mathcal{D}(H_0)}$$
in the Hilbert space $\mathcal{D}(H_0)$ with the norm $\|h\|_{\Gamma(H_0)}$. That means that, for an arbitrary smooth finite function $\psi_2(t)$, the "convolution"
$$\int f(x_1, x_2 - t_2)\, \psi_2(t_2)\, dt_2$$
belongs to $\mathcal{D}(H_0)$ as a function of the variable x_1 for all x_2.

STEP 2. It will be verified that the group of translations $f_0(x_1) \to f_0(x_1 - t)$ is strongly continuous in the space $\mathcal{D}(H_0)$ equipped with the norm $\|h\|_{\Gamma(H_0)}$. That is, it will be shown that
$$\lim_{\epsilon \to 0} \int \|(-i \sigma \frac{\partial}{\partial \xi_1} + \gamma)(f_0(\xi_1 + \epsilon) - f_0(\xi_1))\|\, d\xi_1 = 0 \ .$$
On passage to Fourier transforms, the last identity takes the form
$$\lim_{\epsilon \to 0} \int |e^{i\epsilon\lambda} - 1|^2 (\|(\sigma\lambda + \gamma)\, \tilde{f}_0(\lambda)\|_E^2 + \|\tilde{f}_0(\lambda)\|_E^2)\, d\lambda = 0 \ .$$
For its proof it is sufficient to make use of the inequality
$$\int (\|(\sigma\lambda + \gamma)\, \tilde{f}\|_E^2 + \|\tilde{f}\|_E^2)\, d\lambda < \infty \ ,$$
which is valid because $f_0(\xi_1) \in \mathcal{D}(H_0)$.

STEP 3. Thus $F(\xi_1, \xi_2)$ is a finite smooth function which belongs to $\mathcal{D}(H_0)$ for all ξ_2. It will be shown that $F(\xi_1, \xi_2) \in \mathcal{D}(\gamma)$ for all (ξ_1, ξ_2). Indeed
$$F(\xi_1, \xi_2) = \int_{-\infty}^{\xi_1} \frac{\partial F(\xi_1', \xi_2)}{\partial \xi_1'}\, d\xi_1' \tag{16}$$
and $\frac{\partial F}{\partial \xi_1'}$ is a smooth function of the variable ξ_1' which belongs to $\mathcal{D}(H_0)$. That means that
$$\int ((1 + \gamma^2)\, \frac{\partial F(\xi_1, \xi_2)}{\partial \xi_1},\, \frac{\partial F(\xi_1, \xi_2)}{\partial \xi_1})_E\, d\xi_1 < \infty$$
and because of formula (16), $F(\xi_1, \xi_2) \in \mathcal{D}(\gamma)$ for all ξ_1, ξ_2.

STEP 4. The type of the entire operator-function
$$\exp i\xi_2(z\sigma + \gamma)$$
does not exceed $\|\sigma\| \xi_2$. That means that the function $f(\xi_1,\xi_2)$, considered as a function of ξ_1 for fixed ξ_2, vanishes outside of a $(\|\sigma\| \xi_2)$-neighborhood of the set on which $f_0(\xi_1)$ is nonzero. Therefore, when ψ_1, ψ_2 are chosen to vanish outside of a small neighborhood of the origin, it is possible to make the function $F(\xi_1,0)$ vanish outside of an ϵ-neighborhood of the set where the function $f_0(\xi_1)$ is nonzero. The usual considerations lead to the inequality
$$\int_{-\infty}^{+\infty} \|F(\xi,0) - f_0(\xi_1)\|_E^2 \, d\xi_1 < \infty \, .$$

The first assertion of Theorem 2 has been verified.

Consider now unitarily equivalent representations
$$J_{\sigma_1',\sigma_2',\gamma'} \oplus U' \, , \qquad J_{\sigma_1'',\sigma_2'',\gamma''} \oplus U'' \, .$$
Unitary equivalence of the representations
$$\exp i(zt_1 + z\sigma't_2 + \gamma't_2) \oplus U'(t_1,t_2),$$
$$\exp i(zt_1 + z\sigma''t_2 + \gamma''t_2) \oplus U''(t_1,t_2)$$
is obtained when the vector spaces E', E" are equipped with Hilbert metrics σ_1', σ_1'' and operators σ', γ', σ'', γ'' are defined by the formulas
$$(\sigma'h,h)_E = \sigma_2'(h,h) \, , \qquad (\gamma'h,h)_E = \gamma'(h,h) \, ,$$
$$(\sigma''h,h)_E = \sigma_2''(h,h) \, , \qquad (\gamma''h,h)_E = \gamma''(h,h) \, .$$
But the operator which realizes the unitary equivalence of the one-parameter semigroups of isometries
$$\exp izt_1$$
in the spaces $H_+^2(E')$, $H_+^2(E'')$ is, as is well known, the operator of multiplication by a unitary constant. Let
$$U_0 : E' \to E''$$
be this constant. Then
$$U_0 \exp it_2(z\sigma' + \gamma') = \exp it_2(z\sigma'' + \gamma'') U_0$$
for all t_2, z, Im $z > 0$. That means that
$$U_0 (z\sigma' + \gamma') = (z\sigma'' + \gamma'') U_0 \, ,$$
$$U_0 \sigma' = \sigma'' U_0 \, , \qquad U_0 \gamma' = \gamma'' U_0 \, .$$
That is the operator U_0 is an isomorphism of the spaces E', E" which satisfies the requirements of Theorem 2. This completes the proof of the theorem.

CHAPTER 2

ON ISOMETRIC REPRESENTATIONS OF RANK OF NONUNITARITY THREE

Assume given three nonzero real numbers α_1, α_2, α_3 whose sum is zero. Consider the sets

$$\hat{\Gamma} = \{(z_1, z_2) : (z_2 - \alpha_1)(z_2 - \alpha_2)(z_2 - \alpha_3) - z_1^2 z_2 = 0\},$$

$$\Gamma_+ = \{(z_1, z_2) : (z_1, z_2) \in \hat{\Gamma},\ \text{Im}\ z_1 > 0\},$$

$$\Gamma = \{(z_1, z_2) : (z_1, z_2) \in \hat{\Gamma},\ \text{Im}\ z_1 = 0\}.$$

The aim is to describe within unitary equivalence all completely nonunitary isometric representations $I(x_1, x_2)$, $|x_2| < x_1$ of rank of nonunitarity three for which

$$\text{spset}\ I(x_1, x_2) \subset \Gamma_+$$

and to describe the isometric morphisms

$$i: I(x_1, x_2) \to I_2(x_1, x_2)$$

of such representations.

It is essential that the class so introduced contains, for every representation, all its nontrivial subrepresentations. Note also that the algebraic curve Γ is irreducible. That means that for the relevant representations

$$\text{spset}\ I(x_1, x_2) = \Gamma_+.$$

The representation $I(x_1, x_2)$ is unitarily equivalent to a representation in the space of solutions of the hyperbolic system of equations

$$\{\sigma_2 (-i \frac{\partial}{\partial x_1}) - \sigma_1 (-i \frac{\partial}{\partial x_2} + \gamma)\} u = 0 \qquad (17)$$

where the matrix σ_1 of order three can be taken to be unity, so that

$$\det(\sigma_2 z_1 - \sigma_1 z_2 + \gamma) = \det(\sigma_2 z_1 - z_2 + \gamma)$$
$$= z_1^2 z_2 - (z_2 - \alpha_1)(z_2 - \alpha_2)(z_2 - \alpha_3). \qquad (18)$$

Since the monomials z_1, z_2^2, and z_1^3 are absent on the right side of equation (18) and since the coefficient of $z_1^2 z_2$ is equal to $+1$, the spectrum of the operator σ_2 is found to be

$$\text{spec}\ \sigma_2 = \{-1, 0, 1\}.$$

On the choice of a basis in ∂I, it can be assumed that

$$\sigma_2 = \text{diag}\{-1, 0, 1\}.$$

Since monomials of degree two are absent in formula (18), the diagonal elements of the matrix γ are zero. On the choice of a basis in ∂I, all matrix elements of γ, with the possible exception of the corner ones, are nonnegative. Since the monomial z_1 is absent on the right side of formula (18), the elements of the minor diagonal of the matrix γ are equal. It follows that

$$\gamma = \begin{pmatrix} 0 & a_1 & a_2 \\ a_1 & 0 & a_1 \\ \bar{a}_2 & a_1 & 0 \end{pmatrix}, \quad a_1 \geq 0$$

$$\det\left(\begin{pmatrix} -1 & 0 & 0 \\ 0 & 0 & 0 \\ 0 & 0 & 1 \end{pmatrix} z_1 - z_2 + \begin{pmatrix} 0 & a_1 & a_2 \\ a_1 & 0 & a_1 \\ \bar{a}_2 & a_1 & 0 \end{pmatrix}\right)$$

$$= z_1^2 z_2 - (z_2 - \alpha_1)(z_2 - \alpha_2)(z_3 - \alpha_2) .$$

The system of equations for a_1, a_2 is of the form

$$\begin{cases} 2 \operatorname{Re} a_2 a_1^2 = \alpha_1 \alpha_2 \alpha_3 \\ 2 a_1^2 + |a_2|^2 = -\alpha_1 \alpha_2 - \alpha_2 \alpha_3 - \alpha_1 \alpha_3 \end{cases} \quad (19)$$

Thus every isometric representation of the class under consideration is unitarily equivalent to a representation I_{a_1, a_2} by means of translations in the space of solutions of finite energy in the region $|x_2| < x_1$ of one, and only one, of the following hyperbolic systems of equations

$$i \frac{\partial u_1}{\partial x_1} + i \frac{\partial u_1}{\partial x_2} + a_1 u_2 + a_2 u_3 = 0$$

$$i \frac{\partial u_2}{\partial x_2} + a_1 u_1 + a_1 u_3 = 0 \quad (20)$$

$$-i \frac{\partial u_3}{\partial x_1} + i \frac{\partial u_3}{\partial x_2} + \bar{a}_2 u_1 + a_1 u_2 = 0 .$$

Here $a_1 > 0$ and a_1, a_2 satisfy equations (19).

Another convenient realization of the representation I_{a_1, a_2} is obtained on passage from solutions of the system (17) to their restrictions to the line $x_2 = 0$ and to their Fourier transforms. In such a realization the representation space is $H_+^2 \oplus H_+^2 \oplus H_+^2$ and the operators are the operators of multiplication by

$$\exp i(z_1 t_1 + \begin{pmatrix} z_1 & a_1 & a_2 \\ a_1 & 0 & a_1 \\ \bar{a}_2 & a_1 & z_1 \end{pmatrix} t_2) .$$

The Riemann surface Γ_+ forms a "triple covering" over the half-plane $\operatorname{Im} z_1 > 0$. In a neighborhood of every branch point z_{1*} a "coordinate" τ on Γ_+ is of the form

$$\tau \sim \sqrt{z_1 - z_{1*}} .$$

It will be shown that the representation I_{a_1, a_2} is regular. It is sufficient to show that the operator-function

$$\varphi(z_1, \begin{pmatrix} -z_1 & a_1 & a_2 \\ a_1 & 0 & a_1 \\ \bar{a}_2 & a_1 & z_1 \end{pmatrix})$$

is bounded at branch points $z_{1*} \in \Gamma_+$ for every holomorphic function φ which is bounded on Γ_+. It remains only to make use of the fact that the asymtotic identity

$$\psi(z_1) = \psi_0 + \sqrt{z_1 - z_{1*}} \, \psi_1 + O(\sqrt{z_1 - z_{1*}})$$

applies to every eigenvector of the matrix

$$\begin{pmatrix} -z_1 & a_1 & a_2 \\ a_1 & 0 & a_1 \\ \bar{a}_2 & a_1 & z_1 \end{pmatrix}$$

which undergoes branching in the neighborhood of the point z_{1*}. The coefficients ψ_0 and ψ_1 are obtained as solutions of the equations

$$\begin{pmatrix} -z_1 - z_{1*} & a_1 & a_2 \\ a_1 & -z_{1*} & a_1 \\ \bar{a}_2 & a_1 & z_1 - z_{1*} \end{pmatrix} \psi_0 = 0,$$

$$\begin{pmatrix} -z_1 - z_{1*} & a_1 & a_2 \\ a_1 & -z_{1*} & a_1 \\ \bar{a}_2 & a_1 & z_1 - z_{1*} \end{pmatrix} \psi_1 = \psi_0.$$

The existence of linearly independent solutions ψ_1, ψ_0 of this system follows from the existence of a pole of the eigenprojections of the matrix-function at each branch point [13, p. 98].

The isometric morphisms of the representations under consideration will now be described. This allows a description of the structure of the common invariant subspaces of the operators $I_{a_1, a_2}(t_1, t_2)$.

First note that the set Γ is equipped in a natural way with the structure of a compact Riemann surface of the first kind which forms a double covering of the z_2-plane.

Here the function

$$\text{Im } z_1 = \text{Im}(\frac{\sqrt{z_2(z_2 - \alpha_1)(z_2 - \alpha_2)(z_3 - \alpha_3)}}{z_2})$$

changes sign only on the "cuts" $(-\infty, \alpha_1]$, $[\alpha_1, \alpha_2]$, $[\alpha_3, \infty)$. That means that Γ_+ is one of the leaves of the Riemann surface Γ.

The differential
$$\frac{dz_2}{\sqrt{z_2(z_2-\alpha_1)(z_2-\alpha_2)(z_2-\alpha_3)}}$$
is an abelian differential of the first kind on Γ. That means that the Riemann surface Γ is isomorphic
$$(z_1, z_2) \to \lambda = \int_0^{z_1} \frac{d\xi}{\sqrt{\xi(\xi-\alpha_1)(\xi-\alpha_2)(\xi-\alpha_3)}} \qquad (21)$$
to the torus
$$C/\{z_{mn}: z_{mn} = 2\omega_1 m + 2\omega_2 n\}$$
where
$$2\omega_1 = \int_{C_1} \frac{d\xi}{\sqrt{\xi(\xi-\alpha_1)(\xi-\alpha_2)(\xi-\alpha_3)}} \quad,$$
$$2\omega_2 = \int_{C_2} \frac{d\xi}{\sqrt{\xi(\xi-\alpha_1)(\xi-\alpha_2)(\xi-\alpha_3)}} \quad.$$

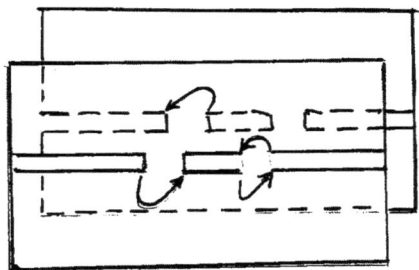

On inverting the elliptic integral (21), it is possible to "solve for" $z_2(\lambda)$, and
$$z_1(\lambda) = \frac{1}{z_2} \frac{dz_2}{d\lambda} \quad.$$
It is easily shown that the region $\operatorname{Im} z_1 > 0$ passes over under the isomorphism (21) into
$$\{\lambda: 0 < \operatorname{Im} \lambda < \omega_2\}/2\omega_1 Z \quad.$$
In other words the strip $\{\lambda: 0 < \operatorname{Im} \lambda < \omega_2\}$ is the universal covering space of the Riemann surface Γ_+.

According to Theorem 3 of Chapter 1 and the above description of the universal covering space of the Riemann surface spset I_{a_1, a_2}, a number τ exists, $0 \le \tau \le 2\pi$, such that the representation I_{a_1, a_2} is unitarily equivalent to the representation
$$f \to \exp i(z_1(\lambda) t_1 + z_2(\lambda) t_2) \, f$$

in the Hilbert space of functions, holomorphic in the strip, satisfying the condition of quasi-periodicity:

$$f(\lambda + 2\omega_1) = e^{i\tau} f(\lambda) .$$

Here

$$\|f\|^2 = \int_{-\omega_1}^{+\omega_1} |f|^2 dz_1 + \int_{-\omega_1+i\omega_2}^{+\omega_1+i\omega_2} |f|^2 dz_1 .$$

Let U be the operator realizing this unitary equivalence:

$$\exp i(z_1(\lambda)t_1 + z_2(\lambda)t_2) U = U I_{a_1,a_2}(t_1,t_2) . \tag{22}$$

With every interior point of the strip $0 < \text{Im } \lambda < \omega_2$ is associated a linear functional δ_λ: $\varphi \to \varphi(\lambda)$ on the space of functions which are holomorphic in the strip. A holomorphic vector-function

$$\psi(x_1,x_2,\lambda) = \frac{1}{\sqrt{2\pi}} \begin{pmatrix} \overline{\psi}_1(x_1,x_2,\overline{\lambda}) \\ \overline{\psi}_2(x_1,x_2,\overline{\lambda}) \\ \overline{\psi}_3(x_1,x_2,\overline{\lambda}) \end{pmatrix} ,$$

is obtained in the strip $-\omega_2 < \text{Im } \lambda < 0$ on passage to $U^* \delta_\lambda$. It satisfied the identity

$$(U u)(\lambda) = \int_0^\infty \sum_{i=1,2} (\sigma_i u, \psi(\overline{\lambda})) \, dx_i \tag{23}$$

for all vectors

$$u = \begin{pmatrix} u_1(x_1,x_2) \\ u_2(x_1,x_2) \\ u_3(x_1,x_2) \end{pmatrix}$$

in the representation space of $I_{a_1,a_2}(t_1,t_2)$.

From (22) and (23) are obtained the identities

$$-i \frac{\partial \psi}{\partial x_1} = z_1(\lambda) \psi, \quad -i \frac{\partial \psi}{\partial x_2} = z_2(\lambda) \psi .$$

Moreover $\psi(x_1,x_2,\lambda)$ is a solution of the hyperbolic system (20). That means that

$$\overline{\psi} = \frac{1}{\sqrt{2\pi}} \begin{pmatrix} \psi_1(\overline{\lambda}) \\ \psi_2(\overline{\lambda}) \\ \psi_2(\overline{\lambda}) \end{pmatrix} \exp i(z_1(\overline{\lambda})x_1 + z_2(\overline{\lambda})x_2) .$$

Note the proof of the following assertions for the vector-functions

$$\overline{\psi}(\overline{\lambda}) = \begin{pmatrix} \psi_1(\lambda) \\ \psi_2(\lambda) \\ \psi_3(\lambda) \end{pmatrix}$$

which are holomorphic in the strip $0 < \text{Im } \lambda < \omega_2$.

1) There exist $\epsilon > 0$, $C > 0$, such that

$$\epsilon < |\psi_1(\lambda)|^2 + |\psi_2(\lambda)|^2 + |\psi_3(\lambda)|^2 \le C . \tag{24}$$

2) The boundary values of the function
$$\sum_{1}^{3} |\psi_j(\lambda)|^2$$
are equal to one almost everywhere on the boundary of the strip $0 < \text{Im } \lambda < \omega_2$.

3)
$$\begin{pmatrix} -z_1(\lambda) & a_1 & a_2 \\ \overline{a}_1 & 0 & a_1 \\ \overline{a}_2 & a_1 & z_1(\lambda) \end{pmatrix} \begin{pmatrix} \psi_1(\lambda) \\ \psi_2(\lambda) \\ \psi_3(\lambda) \end{pmatrix} = z_2(\lambda) \begin{pmatrix} \psi_1(\lambda) \\ \psi_2(\lambda) \\ \psi_3(\lambda) \end{pmatrix}. \qquad (25)$$

Inequality (24) can be written in the form
$$|\psi_1(\lambda)|^2 + |\psi_2(\lambda)|^2 + |\psi_3(\lambda)|^2 \sim 1.$$

It is clear that
$$(|\psi_1(\lambda)|^2 + |\psi_2(\lambda)|^2 + |\psi_3(\lambda)|^2) \frac{1}{2 \text{ Im } z_1(\lambda)} = \|\delta_\lambda\|^2.$$

That means that it is sufficient to prove that
$$\|\delta_\lambda\|^2 \sim \frac{1}{\text{Im } z_1(\lambda)}.$$

Since the function
$$\sum_{1}^{3} |\psi_j(\lambda)|^2$$
is $2\omega_1$-periodic, it is sufficient to establish that
$$\|\delta_\lambda\|^2 \sim \frac{1}{\text{Im } z_1(\lambda)}$$
in the neighborhood of every boundary point of the strip $0 < \text{Im } \lambda < \omega_2$ and to make use of the compactness of the set
$$\{\lambda: 0 \leq \text{Im } \lambda \leq \omega_2\}/\omega_1 Z.$$

Let λ_* be one of the boundary points. For definiteness it can be be assumed that $\text{Im } \lambda_* = 0$. The case $\text{Im } \lambda_* = \omega_2$ is completely analogous.

If the function $z_1(\lambda)$ is holomorphic at the point λ_*, then
$$\frac{dz_1}{d\lambda}\bigg|_{z=\lambda_*} \neq 0$$
since $\text{Im } z_1(\lambda) > 0$ in the strip $0 < \text{Im } \lambda < \omega_2$. Therefore
$$dz_1(\lambda) \sim d\lambda, \quad \|\delta_\lambda\|^2 \sim \frac{1}{\text{Im } \lambda} \sim \frac{1}{\text{Im } z_1(\lambda)}$$
near the point λ_*.

If the function $z_1(\lambda)$ has a pole at the point λ_*, then it is simple
$$z_1(\lambda) \sim \frac{1}{\lambda - \lambda_*}, \qquad (26)$$
since $\text{Im } z_1(\lambda) > 0$ in the strip $0 < \text{Im } \lambda < \omega_2$. From (26) follows the relation

$$\frac{1}{\operatorname{Im} z_1(\lambda)} \sim \frac{1}{\operatorname{Im}(\lambda - \lambda_*)^{-1}} \sim \frac{|\lambda - \lambda_*|^2}{\operatorname{Im} \lambda} \ . \tag{27}$$

But according to the definition introduced in §3 of Chapter 1, the Hilbert space there considered of $2\omega_1$-periodic functions holomorphic in the strip passes over into a "customary Hardy space" under the transformation

$$f(\lambda) \to (z_1(\lambda) + i) \ f(\lambda) \ .$$

Therefore

$$\|\delta_\lambda\|^2 \sim \frac{1}{1 + |z_1(\lambda)|^2} \frac{1}{\operatorname{Im} \lambda} \sim \frac{|\lambda - \lambda_*|^2}{\operatorname{Im} \lambda} \tag{28}$$

Inequality (24) now follows from (27) and (28) in the case of a pole of the function $z_1(\lambda)$ at the point λ_*, and accordingly in the general case.

Equality to one of the boundary values of

$$|\psi_1(\lambda)|^2 + |\psi_2(\lambda)|^2 + |\psi_3(\lambda)|^2$$

follows from the isometric property of the operator U:

$$\int \sum_{1}^{3} |u_j(\xi)|^2 \ d\xi = (\int_{-\omega_1}^{\omega_1} + \int_{-\omega_1 + i\omega_2}^{+\omega_1 + i\omega_2}) \ |\hat{u}_j(\lambda)|^2 \ dz_1 \ ,$$

where

$$\hat{u}_j(\lambda) = \frac{1}{\sqrt{2\pi}} \sum_{1}^{3} \psi_j(\lambda) \int_0^\infty u_j(\xi) \ e^{iz_1(\lambda)\xi} \ d\xi$$

Equation (25) holds because $\psi(x_1, x_2, \lambda)$ is a solution of the system of equations (20).

Properties 1) - 3) of the functions $\psi_j(\lambda)$ defined above by the identity

$$\left(U \begin{pmatrix} u_1 \\ u_2 \\ u_3 \end{pmatrix} \right) (\lambda) = \frac{1}{\sqrt{2\pi}} \sum_{1}^{3} \psi_j(\lambda) \int_0^\infty u_j(\xi) \ e^{-iz_1(\lambda)\xi} d\xi$$

have been verified.

REMARK 1. If the vector-functions,

$$\begin{pmatrix} \psi_1'(\lambda) \\ \psi_2'(\lambda) \\ \psi_3'(\lambda) \end{pmatrix} , \begin{pmatrix} \psi_1''(\lambda) \\ \psi_2''(\lambda) \\ \psi_3''(\lambda) \end{pmatrix} ,$$

which are holomorphic in the strip $0 < \operatorname{Im} \lambda < \omega_2$, satisfy conditions 1) - 3), then they differ only in a constant factor of absolute value one. Indeed it follows from 1) - 3) that

$$\begin{pmatrix} \psi_1''(\lambda) \\ \psi_2''(\lambda) \\ \psi_3''(\lambda) \end{pmatrix} = \varphi(\lambda) \begin{pmatrix} \psi_1'(\lambda) \\ \psi_2'(\lambda) \\ \psi_3'(\lambda) \end{pmatrix}$$

where $\varphi(\lambda)$ is a function, which is holomorphic in the strip, such that
$$\sup |\ln|\varphi(\lambda)|| < \infty . \qquad (29)$$

The harmonic function $\ln|\varphi(\int)|$ extends by the reflection principle to a bounded harmonic function in R^2 which vanishes on the real axis. That means $\ln|\varphi| = 0$ identically and that φ is a unitary constant.

In summary the following result was shown.

PROPOSITION 1. Assume that α_1, α_2, α_3 are nonzero real numbers whose sum is zero. If a_1, a_2 satisfy equations (19), then the canonical spectral set of the semigroup of translations $I_{a_1,a_2}(t_1,t_2)$ in the space of solutions of the system (20), considered with the energy metric, coincides with

$$\Gamma_+ = \{(z_1,z_2): (z_2-\alpha_1)(z_2-\alpha_2)(z_2-\alpha_3) - z_1^2 z_2 = 0, \text{ Im } z_1 > 0\} .$$

Representations $I_{a_1',a_2'}(t_1,t_2)$ and $I_{a_1'',a_2''}(t_1,t_2)$ are not unitarily equivalent.

PROPOSITION 2. Every completely nonunitary isometric representation $I(t_1,t_2)$ of the additive semigroup

$$\{(t_1,t_2): |t_2| < t_1\} ,$$

whose canonical spectral set is contained in the set Γ_+, is unitarily equivalent to one of the representations $I_{a_1,a_2}(t_1,t_2)$.

Here a_1, a_2 satisfy equations (19).

As a corollary of Proposition 2, every subrepresentation of the representation $I_{a_1,a_2}(t_1,t_2)$ is unitarily equivalent to a representation $I_{a_1',a_2'}(t_1,t_2)$. Here a_1', a_2' satisfy equations (19).

Recall that the strip $0 < \text{Im } \lambda < \omega_2$ is the universal covering space of the Riemann surface Γ_+ and that the coordinate functions $z_1(\lambda)$, $z_2(\lambda)$ are $2\omega_1$-periodic functions in the strip. See formula (21).

PROPOSITION 3. A vector-function

$$\begin{pmatrix} \psi_1(a_1,a_2,\lambda) \\ \psi_2(a_1,a_2,\lambda) \\ \psi_3(a_1,a_2,\lambda) \end{pmatrix}$$

exists, unique within a constant factor of absoute value one, which is holomorphic in the strip $0 < \text{Im } \lambda < \omega_2$, such that

$$\sum_1^3 |\psi_j(\lambda)|^2 \sim 1 ,$$

$$\lim_{\text{Im } \lambda \to \{0,\omega_2\}} \sum_1^3 |\psi_j(\lambda)|^2 = 1 ,$$

$$\begin{pmatrix} -z_1(\lambda) & a_1 & a_2 \\ a_1 & 0 & a_1 \\ \bar{a}_2 & a_1 & z_1(\lambda) \end{pmatrix} \begin{pmatrix} \psi_1 \\ \psi_2 \\ \psi_3 \end{pmatrix} = z_2(\lambda) \begin{pmatrix} \psi_1 \\ \psi_2 \\ \psi_3 \end{pmatrix}.$$

As a corollary to Proposition 3, the identities

$$\psi_1(a_1,a_2,\lambda + 2\omega_1) = e^{i\tau(a_1,a_2)} \psi_1(a_1,a_2,\lambda)$$
$$\psi_2(a_1,a_2,\lambda + 2\omega_1) = e^{i\tau(a_1,a_2)} \psi_2(a_1,a_2,\lambda)$$
$$\psi_3(a_1,a_2,\lambda + \omega_1) = e^{i\tau(a_1,a_2)} \psi_3(a_1,a_2,\lambda)$$

are satisfied.

The function $\tau(a_1,a_2) \in R/2\pi Z$ is uniquely determined by these equations.

Recall that $H^2(0,\omega_2,\tau)$ is the Hardy space of functions, holomorphic in the strip $0 < \text{Im } \lambda < \omega_2$, which are quasi-periodic,

$$f(\lambda + 2\omega_1) = e^{i\tau} f(\lambda).$$

The space $H^2(dz_1(\lambda),\tau)$ consists of those functions $f(\lambda)$, which are holomorphic in the strip $0 < \text{Im } \lambda < \omega_2$, such that

$$(z_1(\lambda) + i) f(\lambda) \in H^2(0,\omega_2,\tau).$$

A norm is introduced in $H^2(dz_1(\lambda),\tau)$ by means of the identity

$$\|f\|^2 = \int_{-\omega_1}^{+\omega_1} |f(\lambda + i0)|^2 dz_1(\lambda) + \int_{-\omega_1+i\omega_2}^{+\omega_1+i\omega_2} |f(\lambda + i\omega_2 - i0)|^2 dz_1(\lambda).$$

The following proposition, which was previously established, describes the decomposition of solutions of the system (20) in the sector $|x_2| < x_1$, into its exponential solutions

$$\frac{1}{\sqrt{2\pi}} \begin{pmatrix} \psi_1(a_1,a_2,\lambda) \\ \psi_2(a_1,a_2,\lambda) \\ \psi_3(a_1,a_2,\lambda) \end{pmatrix} e^{i(z_1(\lambda)x_1 + z_2(\lambda)x_2)}.$$

PROPOSITION 4. The formula

$$U : \begin{pmatrix} u_1(x_1,x_2) \\ u_2(x_1,x_2) \\ u_3(x_1,x_2) \end{pmatrix} \to \sum_1^3 \frac{1}{\sqrt{2\pi}} \psi_j(a_1,a_2,\lambda) \int_0^\infty u_j(x_1,0) e^{ix_1 z_2(\lambda)} dx_1$$

defines an isometric linear transfomation of the space of solutions of the system (20) in the sector $|x_2| < x_1$, which have finite energy

$$\sum_1^3 \int_0^\infty |u_j|^2 dx_1,$$

onto the space $H^2(dz_1(\lambda),\tau(a_1,a_2,\lambda))$. It is clear that, when $|t_2| < t_1$,

$$U \begin{pmatrix} u_1(x_1-t_1, x_2-t_2) \\ u_2(x_1-t_1, x_2-t_2) \\ u_3(x_1-t_1, x_2-t_2) \end{pmatrix} = e^{i(t_1 z_1(\lambda) + t_2 z_2(\lambda))} U \begin{pmatrix} u_1(x_1, x_2) \\ u_2(x_1, x_2) \\ u_3(x_1, x_2) \end{pmatrix}.$$

Consider a common invariant subspace of the operators $I_{a_1, a_2}(t_1, t_2)$. To every solution $u(x_1, x_2)$ of the hyperbolic system (20), which is of finite energy, is associated its restriction to the real axis,

$$u(x_1, x_2) \rightarrow u(x_1, 0).$$

The representation space $I_{a_1, a_2}(t_1, t_2)$ becomes identified with $L^2(0, \infty) \oplus L^2(0, \infty) \oplus L^2(0, \infty)$. The space L is translation-invariant: $f(x_1 - t) \in L$ whenever $f(x_1) \in L$ and $t > 0$. According to the Beurling-Lax theorem, the subspace L is determined by "its own" inner operator-function $\theta(z)$. The invariance of the subspace L with respect to a two-parameter semigroup of isometries leads, as has been shown, to the relation

$$\begin{pmatrix} -z & a_1 & a_2 \\ a_1 & 0 & a_1 \\ \overline{a_2} & a_1 & z \end{pmatrix} \theta(z) = \theta(z) \begin{pmatrix} -z & a_1' & a_2' \\ a_1' & 0 & a_1' \\ \overline{a_2}' & a_1' & z \end{pmatrix} \quad (30)$$

where a_1', a_2' satisfy equations (19). Furthermore the following is true.

PROPOSITION 5. If a_1', a_2' satisfy equations (19) and if $\varphi(\lambda)$ is an inner function, holomorphic in the strip $0 < \text{Im } \lambda < \omega_2$, such that

$$\varphi(\lambda + 2\omega_1) = \exp i(\tau(a_1', a_2') - \tau(a_1, a_2)) \varphi(\lambda), \quad (31)$$

then the operator-function

$$\theta(z) = \sum_{\{\lambda : z_1(\lambda) = z\}} \varphi(\lambda) \begin{pmatrix} \psi_1(a_1', a_2', \lambda) \\ \psi_2(a_1', a_2', \lambda) \\ \psi_3(a_1', a_2', \lambda) \end{pmatrix}$$
$$\times (\psi_1(a_1, a_2, \lambda), \psi_2(a_1, a_2, \lambda), \psi_3(a_1, a_2, \lambda)), \quad (32)$$

considered only for real z, is the Beurling-Lax operator-function of some common invariant subspace L of the operators $I_{a_1, a_2}(t_1, t_2)$. Here the representations $I_{a_1', a_2'}$ and $I_{a_1, a_2} | L$ are unitarily equivalent. Conversely if L is a common invariant subspace of the operators I_{a_1, a_2} then its characteristic operator function $\theta(z)$ is of the form (32), where $\varphi(\lambda)$ is a function which is holomorphic in the strip and quasi-periodic. See (31).

REMARK 2. At first sight the function $\varphi(\lambda)$ which was associated with an isometric morphism of the representations,

$$i: I_{a_1',a_2'}(t_1,t_2) \to I_{a_1,a_2}(t_1,t_2) ,$$

depends on the choice of one-parameter semigroup. Actually that is not the case. For, because of (30), the operator i^* transforms the one-dimensional Hilbert space

$$L_\lambda : \{h: I_{a_1,a_2}(t_1,t_2)^* h = \exp i(z_1(\lambda) + z_2(\lambda)t_2) h\}$$

into the Hilbert space

$$L_\lambda' = \{h: I_{a_1',a_2'}(t_1,t_2)^* = \exp i(z_1(\lambda)t_1 + z_2(\theta)t_2) h\}$$

and

$$|\varphi(\lambda)| = \|I^*|_{L_\lambda}\| .$$

The function $\varphi(\lambda)$ is uniquely determined by this identity within a constant factor of absolute value one.

Existence and uniqueness, within a constant factor, of the vector-function

$$\begin{pmatrix} \psi_1(a_1,a_2,\lambda) \\ \psi_2(a_1,a_2,\lambda) \\ \psi_3(a_1,a_2,\lambda) \end{pmatrix}$$

were previously verified.

It is not difficult to obtain "explicit" formulas for

$$\psi_1(a_1,a_2,\lambda), \psi_2(a_1,a_2,\lambda), \psi_3(a_1,a_2,\lambda) .$$

See [16, pp. 65-71].

REMARK 3. The search for invariant subspaces which contain a given subspace reduces in the present case to a factorization

$$\varphi(\lambda) = \varphi_1(\lambda) \varphi_2(\lambda)$$

where $\varphi_{1,2}(\lambda)$ are holomorphic functions in the strip $0 < \text{Im } \lambda < \omega_2$ with automorphic moduli,

$$|\varphi_{1,2}(\lambda + 2\omega_1)| = |\varphi_{1,2}(\lambda)| .$$

It is not difficult to obtain multiplicative integral representations for $\varphi(\lambda)$ on previously factoring out "Blaschke factors." See [15].

CHAPTER 3
UNITARY DILATIONS OF MULTI-PARAMETER SEMIGROUPS OF CONTRACTIONS

§1. The Functional Model of Unitary Dilations.

Recall that a unitary dilation of a representation $T(x)$ of the semigroup K in a space H is a unitary representation $U(x)$ of the semigroup in a space $H_0 \supset H$ for which
$$T(x) h = P_H U(x) h$$
for all $x \in K$, $h \in H$.

Here P_H is the orthogonal projection of H_0 onto the space H.

A unitary dilation is said to be minimal if H_0 is the smallest reducing subspace of the operators $U(x)$ which contains H.

The existence and uniqueness of a unitary dilation will be discussed in the following section and in Chapter 4. In the present section is obtained a functional model of a unitary dilation $U(x)$ of a completely nonunitary contractive representation $T(x)$.

The necessary notation will be introduced. The smallest common invariant subspace of the operators $U(x)$, $x \in K$, which contains H, is denoted H_+, and the orthogonal complement of H in H_+ is denoted H_{0+}. The isometric representations $U(x)|_{H_+}$ and $U(x)|_{H_{0+}}$ of the semigroup K are denoted $I_+(x)$ and $I_{0+}(x)$ respectively.

Consider the inclusion operators $i: H_{0+} \to H_+$. This is an intertwining operator for the isometric representations
$$I_{0+}(x), \quad I_+(x) .$$

Because of the uniqueness of the minimal unitary dilation of an isometric representation, it is sufficient to construct a functional model of the morphism
$$i: I_{0+}(x) \to I_+(x) .$$

STEP 1. The representation $I_{0+}(x)$ is completely nonunitary. Indeed if $h \in H_0$ and $\forall x \in K$,
$$\|I_{0+}(x)^* h\| = \|h\| ,$$
then $\forall x \in K$, $U(x)^* h \in H_{0+}$. That means that $\forall x \in R^n$, $U(x) h \in H_{0+}$. Thus the orthogonal complement of the linear span of the vectors $H(x) h$, $x \in R^n$, reduces all operators $U(x)$ and contains H_{0+}. Because of the minimality of the unitary dilation, this orthogonal complement coincides with H_0, and $h = 0$.

STEP 2. Decompose the isometric representation $I_+(x)$ in the space H_+ into the orthogonal sum of a completely nonunitary and a unitary representation:
$$I_+(x) = I_+'(x) \oplus U''(x)$$
in the spaces
$$H_+' = \{h \in H_+: \lim_{|x| \to \infty} I_+(x)^* h = 0\}$$
and H_+'' equal to the orthogonal complement of H_+' in H_+.

Consider more closely the inclusion
$$i_1: I_{0+}(x) \to I_+'(x) \oplus U''(x).$$
Let P' and P" be the orthogonal projections of the space H_+ onto the subspaces H_+', H_+'', and let
$$R(x) = U''(x)|_{P''i\, H_{0+}}.$$

LEMMA 1. The representation $U''(x)$ is the minimal unitary dilation of the isometric representation $R(x)$.

PROOF. In the contrary case a nonzero vector $h \in H_+''$ would be found such that $U(x) h = U''(x) h \perp i\, H_{0+}$ for all $x \in R^n$.
But then $h \in H$, $U(x) h \in H$ for all $x \in R^n$. A contradiction is obtained since the representation $P_H U(x)|_H$ is completely nonunitary.

STEP 3. Thus it is sufficient to construct a functional model of the isometric intertwining operator i_1. But from the construction of the representation $R(x)$, it is clear that the operator i_1 is determined within unitary equivalence by the contractive intertwining operator
$$P'\, i_1: I_{0+}(x) \to I_+'(x).$$

STEP 4. In view of Theorem 1 of Chapter 1, the representations $I_+'(x)$ and $I_{0+}(x)$ are unitarily equivalent to the representations
$$f(z) \to \exp i(zt_1 + z \sum_2^n \sigma_j' t_j + \sum_2^n \gamma_j' t_j) f(z),$$
$$f(z) \to \exp i(zt_1 + z \sum_2^n \sigma_j'' t_j + \sum_2^n \gamma_j'' t_j) f(z)$$
in the Hardy spaces $H_+^2(E')$, $H_+^2(E'')$ of holomorphic vector-functions in the upper half-plane with values in some Hilbert spaces E', E". Here

$$\sigma_j'^* = \sigma_j',\quad \sigma_j''^* = \sigma_j'',\quad \gamma_j'^* = \gamma_j',\quad \gamma_j''^* = \gamma_j'',$$
$$\|\sigma_j'\| < \infty,\quad \|\sigma_j''\| < \infty$$
(33)

and
$$[\sigma_j' t + \gamma_j', \sigma_k' t + \gamma_k'] = 0$$
$$[\sigma_j'' t + \gamma_j'', \sigma_k'' t + \gamma_k''] = 0.$$
(34)

That means that the contractive operator which intertwines the representations $I_+'(x)$, $I_{0+}(x)$ passes over, under the unitary equivalence of Theorem 1, into the operator of multiplication by a contractive operator-function $S(z): E'' \to E'$ which is holomorphic in the upper half-plane.

Here
$$S(z)(\sigma_j'' z + \gamma_j'') = (\sigma_j' z + \gamma_j') S(z) . \qquad (35)$$

The operators
$$\exp i(zt_1 + z \sum_2^n \sigma_j'' t_j + \sum_2^n \gamma_j'' t_j)$$
preserve the seminorm
$$\sqrt{\|h\|^2 - \|S h\|^2} .$$

Indeed
$$\| \exp i(zt_1 + z \sum_2^n \sigma_j'' t_j + \sum_2^n \gamma_j'' t_j) h \| = \|h\|$$
and
$$\|S \exp i(zt_1 + z \sum_2^n \sigma_j'' t_j + \sum_2^n \gamma_j'' t_j) h \|$$
$$= \| \exp i(zt_1 + z \sum_2^n \sigma_j' t_j + \sum_2^n \gamma_j' t_j) S h \| = \|S h\| .$$
That is
$$[\sqrt{I - S^*S}, \exp i(zt_1 + z \sum_2^n \sigma_j'' t_j + \sum_2^n \gamma_j'' t_j)] = 0 .$$

In summary, the triple
$$i: I_{0+}(x) \to I_+(x)$$
and consequently the pair
$$(T(x), U(x)) ,$$
are determined within unitary equivalence by a holomorphic contractive operator-function $\theta(z)$ and the operators
$$\sigma_2', \ldots, \sigma_n', \gamma_2', \ldots, \gamma_n' \text{ in } E' ,$$
$$\sigma_2'', \ldots, \sigma_n'', \gamma_2'', \ldots, \gamma_n'' \text{ in } E'' ,$$
which satisfy equations (33), (34), (35).

The following result has been obtained.

THEOREM 1. Let $T(x): H \to H$ be a completely nonunitary contractive representation of the semigroup K and let $U(x): H_0 \to H_0$ be one of its minimal unitary dilations. Then the representations $T(x)$, $U(x)$ uniquely determine two Hilbert spaces E', E'', two choices of selfadjoint operators
$$\sigma_2', \ldots, \sigma_n', \gamma_2', \ldots, \gamma_n' \text{ in } E',$$
$$\sigma_2'', \ldots, \sigma_n'', \gamma_2'', \ldots, \gamma_n'' \text{ in } E'',$$
and a contractive operator-function $S(z): E'' \to E'$, which is holomorphic in the upper half-plane, satisfying equations (33), (34), and (35).

Here the pair $(T(x), U(x))$ is unitarily equivalent to the following "model" pair of representations:

$$U(t_1, \ldots, t_n): f(z) \oplus g(x)$$
$$\to \exp i(zt_1 + z \sum_2^n \sigma_j' t_j + \sum_2^n \gamma_j' t_j) f(z) \qquad (36)$$
$$\oplus \exp i(xt_1 + x \sum_2^n \sigma_j'' t_j + \sum_2^n \gamma_j'' t_j) g(x),$$

$$T(t_1, \ldots, t_n) = P_{H(S)} U(t_1, \ldots, t_n)|H(S). \qquad (37)$$

Here
$$H(S) = H_{0+}(S) + H_+(S),$$
$$H_0(S) = L^2(E') \oplus L^2(\sqrt{1 - S^*S}\, dx),$$
$$H_+(S) = H_+^2(E') \oplus L^2(\sqrt{1 - S^*S}\, dx),$$
$$H_{0+}(S) = \{Sf \oplus P_{\operatorname{Im} \sqrt{1 - S^*S}} f: f \in H_+^2(E'')\}$$

and $H_{0+}(S)$ is the orthogonal complement of $H_+(S)$ in $H(S)$, and $P_{H(S)}$ is the orthogonal projection of $H_0(S)$ onto the subspace $H(S)$, and

$$P_{\operatorname{Im} \sqrt{1 - S^*S}}$$

is the orthogonal projection of E'' onto the closure of the range of the operator $\sqrt{1 - S^*S}$.

REMARK 1. It is clear that for any E', E'', $\sigma_2', \ldots, \sigma_n', \gamma_2', \ldots, \gamma_n', \sigma_2'', \ldots, \sigma_n'', \gamma_2'', \ldots, \gamma_n''$, S satisfying conditions (33), (34), (35), formulas (36), (37) determine, for small $\|\sigma_j''\|$, $\|\sigma_j'\|$, a contractive representation of the semigroup K and its minimal unitary dilation.

REMARK 2. It is not difficult to obtain the model of the unitary dilation without using distinguished one-parameter semigroups. (The semigroup $(t, 0, \ldots, 0)$ was distinguished in Theorem 1.) Such an "invariant" approach to the construction of the model is described in §2 of Chapter 1.

REMARK 3. On passage from the representations $(T(x), U(x))$ to the adjoint representations $(T(x)^*, U(x)^*)$, the spaces and operators appearing in the statement of Theorem 1 undergo the following transition

$$E' \to E'', \quad E'' \to E', \quad S(z) \to S(-z)^*,$$
$$\sigma_j' \to \sigma_j'', \quad \sigma_j'' \to \sigma_j', \quad \gamma_j' \to -\gamma_j'', \quad \gamma_j'' \to -\gamma_j'.$$

§2. Core Unitary Dilations of Contractive Representations.

It is well known that the minimal unitary dilation of a one-parameter semigroup of contractions is unique within unitary equivalence. Uniqueness of a minimal unitary dilation is lost on passage to multi-parameter semigroups.

A special class of unitary dilations will be distinguished for which uniqueness occurs with semigroups of arbitrary dimensions.

One such class of dilations was studied in the work of Bram, Nagy, and Halperin [1], where these dilations are called regular. It can be shown that the study of regular unitary dilations leads in important special cases to the very difficult problem [3] of describing the common invariant subspaces of operators of multiplication by the independent variables in a Hardy space of functions in a polydisk. Furthermore even among pairs of finite-dimensional operators, some exist which have no regular unitary dilations.

The aim of the present section is the study of a class of unitary dilations other than those of Nagy-Bram. The harmonic analysis of these unitary dilations is related, for a wide class of representations, to problems about invariant subspaces in Hardy spaces on Riemann surfaces, that is essentially to the theory of functions of one complex variable.

First some supplementary information about contractive representations is needed.

In the remainder of §2, only those contractive representations $T(t_1,\ldots,t_n): H \to H$ are considered which admit at least one unitary dilation. The question of existence of a dilation will be considered in §1 of Chapter 4 and in the supplement to Chapter 4.

Consider the infinitessimal operators $\Sigma\, t_j A_j$ of one-parameter semigroups of contractions $T(\tau t_1,\ldots,\tau t_n)$ and the linear span $\mathcal{B}_T$ of vectors of the form
$$\int\cdots\int \varphi(t_1,\ldots,t_n)\, T(t_1,\ldots,t_n)^* h\, dt_1\cdots dt_n ,$$
where $h \in H$, $\varphi(t_1,\ldots,t_n) \in C_0^\infty(R^n)$ and the function f vanishes outside the cone K. It is well known that $\mathcal{B}_T$ is a dense linear subset of K.

The factorization of $\mathcal{B}_T$ by the kernel of the sesquilinear form
$$[h_1,h_2] = -\frac{d}{dt}\Big|_{t=0} (T(t,0,\ldots,0)^* h_1, T(t,0,\ldots,0)^* h_2)$$
produces a vector space $\partial_0 T$, and the completion of $\partial_0 T$ with respect to the norm
$$\sqrt{[h,h]}$$
is the Hilbert space which has been denoted ∂T. The vector space $\partial_0 T$ will be called the boundary lineal and the space ∂T will be called the boundary space of the representation T.

The "operator factactorization" $\mathcal{B}_T \to \partial T$ which arises under the described construction of the boundary space ∂T, is denoted φ_T.

Note that $\varphi_T \mathcal{B}_T = \partial_0 T$ where $\partial_0 T$ is dense in ∂T.

According to Theorem 1 of Chapter 3, a contractive operator
$$j: \mathfrak{H}_T \to L^2((0,\infty), E')$$
exists which intertwines the contractive representation $T^*(t_1, \ldots, t_n)$ and the coisometric representation
$$\exp i(-i \frac{d}{d\xi}(t_1 + \sum_2^n t_j \sigma_j') + \sum_2^n t_j \gamma_j') \; .$$
Then clearly $\forall \, h \in \mathfrak{H}_T$,
$$- \frac{d}{dt}\Big|_{t=0} (\|T(t,0,\ldots,0) h\|^2) = \|j h\|_{\xi=0}\|^2 \; .$$
It is therefore possible to assume that
$$\partial_0 T \subset \partial T \subset E', \quad \varphi_T h = j h \big|_{\xi=0} \; .$$

Recall the definition of a core selfadjoint operator γ with domain $\mathfrak{D}(\gamma)$ in a Hilbert space E. A vector subspace L is said to be a core of γ if
1) $L \subset \mathfrak{D}(\gamma)$,
2) $\overline{L} = E$,
3) the restriction $\gamma_{|L}$ is an essentially selfadjoint operator.

DEFINITION. A minimal unitary dilation $U(t_1, \ldots, t_n)$ of a representation $T(t_1, \ldots, t_n)$ is said to be a core representation if the boundary linear $\partial_0 T \subset E'$ is a core of each of the operators $\gamma_1', \ldots, \gamma_n'$ and if the boundary linear $\partial_0 T^* \subset E''$ of the adjoint representation is a core of each of the operators $\gamma_2'', \ldots, \gamma_n''$.

REMARK 4. The definition of a core representation is correct when the functional model of a unitary dilation, obtained in §1 of Chapter 3, is essentially unique. Proceeding in the spirit of §2 of Chapter 1, it is not difficult to give a definition of a core representation which is equivalent to the one introduced, but which does not make use of the distinguished one-parameter semigroup $(t,0,\ldots,0)$ of the semigroup K.

EXAMPLE 1. Consider in $L^2(0,\infty)$ a pair of commuting maximal dissipative operators
$$-i \frac{d}{dx} \; , \; -i \sigma \frac{d}{dx} + \gamma \; , \quad \sigma \geq 0, \; \gamma \in R \tag{38}$$
defined on the vector space of absolutely continuous functions with square integrable derivative. It is clear that, for every complex number a_1 with positive imaginary part, the functions
$$\text{const } e^{i a_1 x}$$
form a one-dimensional eigensubspace of these operators.

On passage to the differential operators A_1, A_2, defined by formulas (38), and on the introduction of the unitary representation
$$\exp i(A_1 t_1 + A_2 t_2) \; ,$$

one obtains a core unitary dilation of the one-dimensional representation

$$\exp i(a_1 t_1 + a_2 t_2) \, , \quad a_2 = \sigma a_1 + \gamma \, .$$

EXAMPLE 2. Introduce the differential operators

$$-i \frac{d}{dx} \, , \quad -i \frac{d}{dx} + \begin{pmatrix} 0 & -1 \\ 1 & 0 \end{pmatrix}$$

in the space $L^2(0,\infty) \oplus L^2(0,\infty)$.

Let $h_1 e^{iax} \oplus h_2 e^{iax}$ be a common eigenfunction of these operators, Im $a > 0$. Proceeding as in Example 1, one obtains the minimal unitary dilation of a one-dimensional representation in $L^2(-\infty,+\infty) \oplus L^2(-\infty,+\infty)$. In the present case the dimension of the boundary linear is equal to 1, and the dimension of the space E is equal to 2. That means that the unitary dilation is not a core dilation.

EXAMPLE 3. In the previous example, the boundary linear was not dense in the space E'. Construct the dilation, which is not a core representation, for which $\partial_0 T = \partial T = E'$.

Define regions Ω_n for integer n by means of the formula

$$\Omega_n = \{(\xi_1, \xi_2) : |\xi_2 - 2\pi n| < \pi - \xi_1, \, \xi_1 > 0\}$$

and the region $\Omega = \cup \Omega_n$. Fix $\omega \in [0, 2\pi)$.

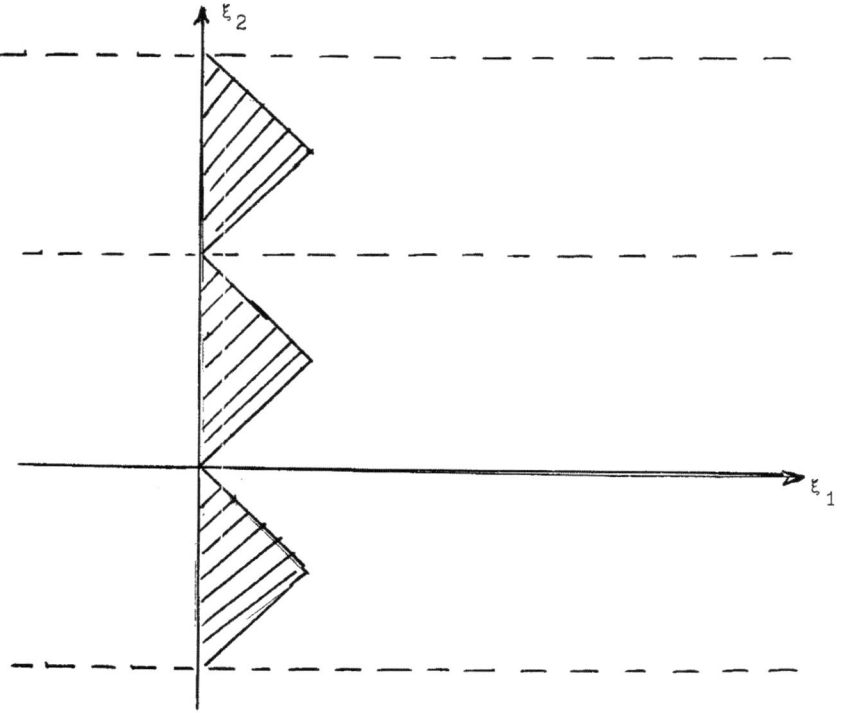

In the space of those locally square-integrable functions $f(\xi_1, \xi_2)$ for which
$$f(\xi_1, \xi_2 + 2\pi) = e^{i\omega} f(\xi_1, \xi_2) ,$$
introduce a norm by
$$\|f\|^2 = \iint |f|^2 \, d\xi_1 d\xi_2 .$$
The Hilbert space obtained is denoted L_ω^2 and the subspace, consisting of those functions which vanish outside of Ω, is denoted H_ω. It is clear that the formula
$$f(\xi_1, \xi_2) \to f(\xi_1 + t_1, \xi_2 + t_2) \chi_\Omega$$
defines a contractive representation τ_ω of the additive semigroup $\{(t_1, t_2) : |t_2| < t_1\}$ in the space H_ω and its unitary dilation U_ω in L_ω^2. It is not difficult to show that the space E' can be identified with $L^2(-\pi, \pi)$ and that the operator $\gamma(\omega)$ thereby coincides with the operator of differentiation, defined on the vector space $\mathcal{D}(\gamma(\omega))$ consisting of those absolutely continuous functions f with derivative in $L^2(-\pi, +\pi)$ such that
$$f(+\pi) = e^{i\omega} f(-\pi) .$$

The domain of the closure of the restricted operator in the boundary space is distinguished from $\mathcal{D}(\gamma(\omega))$ by the boundary conditions $f(+\pi) = f(-\pi) = 0$.

That means that the unitary dilations U_ω are not core representations. Since the representations τ_ω are pairwise unitarily equivalent, an instructive example of nonuniqueness of unitary dilations is obtained.

Analogous Examples 1 and 2 illustrate the nonuniqueness of a minimal unitary dilation of a one-dimensional representation of the semigroup K.

THEOREM 2. *The core unitary dilation of a contractive representation* $T(t_1, \ldots, t_n) : H \to H$ *of the semigroup K is unique within unitary equivalence.*

PROOF. As is seen from the definition of a core representation, it is sufficient to show that the sesquilinear forms
$$((t_1 + \sum_2^n t_k \sigma_k') \varphi_T h, \varphi_T h) ,$$
$$(\sum_2^n t_k \gamma_k' \varphi_T h, \varphi_T h) ,$$
$$((t_1 + \sum t_k \sigma_k'') \varphi_{T*}, \varphi_{T*}) ,$$
$$(\sum_2^n t_k \gamma_k'' \varphi_{T*} h, \varphi_{T*} h)$$
in $\mathcal{D}_T$, $\mathcal{D}_{T*}$ are independent of the choice of unitary dilation. But the identities

$$((t_1 + \sum_2^n t_k \sigma_k') \varphi_T h, \varphi_T h) = -2 \operatorname{Im}(\Sigma\, t_k A_k^* h, h),$$

$$((t_1 + \sum_2^n t_k \sigma_k'') \varphi_{T^*} h, \varphi_{T^*} h) = 2 \operatorname{Im}(\Sigma\, t_k A_k h, h),$$

$$(\sum_2^n t_k \gamma_k' \varphi_T h, \varphi_T h) = \tfrac{1}{i} \{(\sum_2^n t_k A_k^* h, A_1^* h) - (A_1^* h, \sum_2^n t_k A_k^* h)\},$$

$$(\sum_2^n t_k \gamma_k'' \varphi_{T^*} h, \varphi_{T^*} h) = \tfrac{1}{i} \{(A_1 h, \sum_2^n t_k A_k h) - (\sum_2^n t_k A_k h, A_1 h)\}.$$

This completes the proof of Theorem 2.

CHAPTER 4

FUNCTIONAL MODELS OF MULTI-PARAMETER SEMIGROUPS OF CONTRACTIONS

§1. Model Construction within Unitary Equivalence

A simple sufficient condition will be obtained for the existence of a core unitary dilation of a two-parameter semigroup of contractions.

THEOREM 1. If A_1, A_2 are bounded commuting linear operators and if

$$|\operatorname{Im} A_2| \leq \frac{1}{\ell} \operatorname{Im} A_1$$

where $\ell > 0$, then the representation

$$\exp i(A_1 t_1 + A_2 t_2)$$

of the additive semigroup

$$\{(t_1, t_2) : |t_2| < t_1\}$$

admits a core unitary dilation.

PROOF 1. Associate with every vector $h \in H$ a vector-function with values in the subspace

$$G = \overline{\operatorname{Im} A_1 H + \operatorname{Im} A_2 H}$$

by

$$J : h \to P_G \exp \tfrac{1}{i}(A_1^* t_1 + A_2^* t_2) h$$

where P_G is the orthogonal projection of the space H onto G.

It is clear that

$$\frac{\partial (J h)}{\partial t_1} = -i A_1^* J h, \qquad \frac{\partial (J h)}{\partial t_2} = -i A_2^* J h.$$

Therefore

$$\frac{A_2 - A_2^*}{i} \frac{\partial (J h)}{\partial t_1} = (-A_2 A_1^* + A_2^* A_1^*) J h,$$

$$\frac{A_1 - A_1^*}{i} \frac{\partial (J h)}{\partial t_2} = (-A_1 A_2^* + A_1^* A_2^*) J h.$$

On introducing the operators

$$\sigma_1 = \tfrac{1}{i}(A_1 - A_1^*)|_G, \qquad \sigma_2 = \tfrac{1}{i}(A_2 - A_2^*)|_G$$

$$\gamma_{12} = \tfrac{1}{i}(A_2 A_1^* - A_1 A_2^*)|_G,$$

one obtains

$$i \sigma_2 \frac{\partial (J h)}{\partial t_1} - i \sigma_1 \frac{\partial (J h)}{\partial t_2} + \gamma_{12}(J h) = 0. \tag{39}$$

It will be shown that the vector-function $J h$ belongs to the space of the representation $I_{\sigma_1, \sigma_2, \gamma_{12}}$ which was introduced in the

first chapter. Strictly speaking, the subspace G needs to be included
$G \to E$ in its completion with respect to the norm defined by
$$\|h\|^2 = (\text{Im } A_1 h, h) .$$
Since the operator J is a contraction,
$$\|h\|^2 \geq \int_0^\infty \sum_i (\sigma_i \, J \, h, H \, h) \, dt_i ,$$
the inclusion
$$J \, h \subset H_{+\sigma_1, \sigma_2, \gamma_{12}}$$
needs to be verified only for a dense of vectors h. Solve the Cauchy problem
$$(i \, \sigma_2 \frac{\partial}{\partial t_1} - i \, \sigma_1 \frac{\partial}{\partial t_2} + \gamma_{12}) \, u = 0$$
in the sector K with the initial data
$$u(t_1, 0) = (J \, h)(t_1, 0)$$
on the right half-plane, as described in the first part of this paper. For a dense linear space of vectors h, one obtains a smooth solution, belonging to the space $H_{+\sigma_1, \sigma_2, \gamma_{12}}$ which coincides with $(J \, h)(t_1, 0)$ on the half-line $t_2 = 0$, $t_1 > 0$. Needless to say, a smooth solution of the Cauchy problem for the hyperbolic system (39) is unique. That means that $J \, h \in H_{+\sigma_1, \sigma_2, \gamma_{12}}$. It is clear that the operator J is a contraction and that
$$(J \, \exp(-i(A_1^* \, a_1 + A_2^* \, a_2)) \, h)(t_1, t_2) = J \, h(t_1 + a_1, t_2 + a_2) .$$

The remainder of the construction of the unitary dilation is a completely standard procedure: Factor the space H with respect to the kernel of the quadratic form
$$\|h\|^2 - \|J \, h\|^2$$
and complete with respect to the corresponding norm. Denote the resulting Hilbert space by H_R and the operator of factoring and inclusion $H \to H_R$ by J_R. The results of Chapter 1 apply to the two-parameter semigroup of isometries appearing in the space H_R. This semigroup of isometries is unitarily equivalent to the semigroup of shift operators $I_{+\sigma_1', \sigma_2', \gamma_{12}'}$. It is permissible to assume that the representation
$$\exp i(A_1 t_1 + A_2 t_2)$$
is completely nonunitary. The operator
$$J \oplus J_R : H \to H_{\sigma_1, \sigma_2, \gamma_{12}} \oplus R$$
provides the isometric dilation of the two-parameter semigroup of contractions
$$\exp i(A_1 t_1 + A_2 t_2) .$$

The unitary dilation of an isometric representation of a semigroup K is, as has been shown, essentially unique.

Because of the boundedness of the operator γ_{12}, it follows immediately from the definitions that the constructed dilation is a core representation.

REMARK 1. In Section 1 of Chapter 3, a fundamental model of a unitary dilation was constructed. In the case of two-parameter semigroups of contractions, it is convenient to change the notation. Instead of σ_2', σ_2'', γ_2', γ_2'', write σ', σ'', γ', γ''. The transition from the solutions $u(t_1, t_2)$ of the hyperbolic system of equations to vector-functions which are holomorphic in the upper half-plane is realized with the help of the unitary operator

$$u(t_1, t_2) \to \frac{1}{\sqrt{2\pi}} \sqrt{A_1 - A_1^*} \int_0^\infty u(t_1, 0) \, e^{-izt_1} \, dt_1 \, .$$

Here

$$\sigma' = |A_1 - A_1^*|^{-\frac{1}{2}} \frac{A_2 - A_2^*}{i} |A_1 - A_1^*|^{-\frac{1}{2}} \, , \quad (40)$$

$$\gamma' = |A_1 - A_1^*|^{-\frac{1}{2}} \frac{(A_1 - A_1^*) A_2^* - (A_2 - A_2^*) A_1^*}{i} |A_1 - A_1^*| \quad (41)$$

From the theory of unitary dilations of one-parameter semigroups of contractions in the case $\|A_1\| < \infty$, it is well-known that it is possible to choose $E' = E''$ in the functional model of a unitary dilation (§1) and then

$$S(z) = I - (A_1 - A_1^*) \frac{1}{z} + O\left(\frac{1}{z^2}\right) \, .$$

It is not difficult to show that $\sigma'' = \sigma'$,

$$\gamma'' = |A_1 - A_1^*|^{-\frac{1}{2}} \frac{(A_1 - A_1^*) A_2 - (A_2 - A_2^*) A_1}{i} |A_1 - A_1^*|^{-\frac{1}{2}} \, .$$

For the proof it is sufficient to consider the Laurent expansion at infinity of both sides of the equation

$$S(z) (\sigma'' z + \gamma'') = (\sigma' z + \gamma') S(z)$$

and equate coefficients of z^1, z^0. The second equation is equivalent to the evident, in view of (40), relation

$$\gamma'' - \gamma' = |A_1 - A_1^*|^{-\frac{1}{2}} \frac{(A_1 - A_1^*)(A_2 - A_2^*) - (A_2 - A_2^*)(A_1 - A_1^*)}{i} |A_1 - A_1^*|^{-\frac{1}{2}} \, .$$

Thus $\gamma'' = \gamma'$ if, and only if, the operators Im A_1 and Im A_2 commute.

Recall that a pair of operators A_1, A_2 in H is said to be completely nonselfadjoint if no nontrivial common reducing subspace exists to which the restriction of the operators is selfadjoint.

COROLLARY 1 TO THEOREM 1. Let A_1 be a completely nonselfadjoint bounded dissipative operator in a Hilbert space H with characteristic operator-function
$$S(z): E \to E$$
where
$$E = \overline{(A_1 - A_1^*) H} .$$
If $A_2, \ldots, A_n$ are commuting bounded operators in H and if
$$|\operatorname{Im} A_j| \leq \text{const} \times \operatorname{Im} A_1 ,$$
then the family of operators
$$\exp i \Sigma A_j t_j$$
is unitarily equivalent to the following family of operators in the space $H(S)$:
$$f_1(z) \oplus f_2(z)$$
$$\to P_{H(S)} \exp i(zt_1 + z \sum_2^n \sigma_j t_j + \sum_2^n \gamma_j' t_j) f_1(z)$$
$$\oplus \exp i(xt_1 + x \sum_2^n \sigma_j t_j + \sum_2^n \gamma_j'' t_j) f_2(x)$$
where $H(S)$ is the orthogonal complement of
$$\{S f \oplus P_{\operatorname{Im} \sqrt{1-S^*S}} f: f \in H_+^2(\partial H)\}$$
in
$$H_+^2(\partial H) \oplus L^2(\sqrt{1-S^*S} \, dx) ,$$
$P_{H(S)}$, $P_{\operatorname{Im} \sqrt{1-S^*S}}$ are the orthogonal projections onto $H(S)$ and onto the closure of the range of $\sqrt{1-S^*S}$ in ∂H,

$$\sigma_j = |A_1 - A_1^*|_E^{-\frac{1}{2}} \frac{A_j - A_j^*}{i} |A_1 - A_1^*|_E^{-\frac{1}{2}} ,$$

$$\gamma_j' = |A_1 - A_1^*|_E^{-\frac{1}{2}} \frac{(A_1 - A_1^*) A_j^* - (A_j - A_j^*) A_1^*}{i} |A_1 - A_1^*|_E^{-\frac{1}{2}} ,$$

$$\gamma_j'' = |A_1 - A_1^*|_E^{-\frac{1}{2}} \frac{(A_1 - A_1^*) A_j - (A_j - A_j^*) A_1}{i} |A_1 - A_1^*|_E^{-\frac{1}{2}} .$$

The corollary, together with Theorem 2 and equation (3) in the introduction to the present manuscript, leads to the following criterion for the existence of a unitary dilation.

THEOREM 2. If $A_1,\ldots,A_n$ are bounded commuting linear operators in H such that
$$|\operatorname{Im} A_j| \leq \operatorname{const} \times \operatorname{Im} A_1,$$
then the following conditions are equivalent:

1) A core dilation of the representation
$$\exp i \Sigma A_j t_j$$
exists.

2) The identities
$$[\operatorname{Im} A_j \ z + (\operatorname{Im} A_1 \ A_j^* - \operatorname{Im} A_j \ A_1^*)] \ (\operatorname{Im} A_{1|E})^{-1}$$
$$\times [\operatorname{Im} A_k \ z + (\operatorname{Im} A_1 \ A_k^* - \operatorname{Im} A_k \ A_1^*)]_{|E}$$
$$= [\operatorname{Im} A_k \ z + (\operatorname{Im} A_1 \ A_k^* - \operatorname{Im} A_k \ A_1^*)] \ (\operatorname{Im} A_{1|E})^{-1}$$
$$\times [\operatorname{Im} A_j \ z + (\operatorname{Im} A_1 \ A_j^* - \operatorname{Im} A_j \ A_1^*)]_{|E}$$
hold for $j,k = 1,2,\ldots,n$.

3) The identities
$$[\operatorname{Im} A_j \ z + (\operatorname{Im} A_1 \ A_j - \operatorname{Im} A_j \ A_1)]_{|E} \ (\operatorname{Im} A_{1|E})^{-1}$$
$$\times [\operatorname{Im} A_k \ z + (\operatorname{Im} A_1 \ A_k - \operatorname{Im} A_k \ A_1)]_{|E}$$
$$= [\operatorname{Im} A_k \ z + (\operatorname{Im} A_1 \ A_k - \operatorname{Im} A_k \ A_1)] \ (\operatorname{Im} A_{1|E})^{-1}$$
$$\times [\operatorname{Im} A_j \ z + (\operatorname{Im} A_1 \ A_j - \operatorname{Im} A_j \ A_1)]_{|E}$$
hold for $j,k = 1,2,\ldots,n$.

The conditions of Theorem 1 admit a simple interpretation in the special case that
$$\lim_{t \to \infty} \exp(-iA_1^* t) h = 0$$
for all h in H. As was shown in the proof of Theorem 1 (see also §2 of Chapter 1), the space H can be isometrically embedded into the space of solutions of a hyperbolic system of partial differential equations with constant coefficients,
$$i \ \sigma_k \frac{\partial u}{\partial x_j} - i \ \sigma_j \frac{\partial u}{\partial x_k} + \gamma_{jk} \ u = 0$$
where
$$\sigma_k = \frac{A_k - A_k^*}{i}, \quad \gamma_{kj} = \frac{(A_k - A_k^*) \ A_j^* - (A_j - A_j^*) \ A_k^*}{i}.$$

In the case of more than two variables, such a hyperbolic system is ordinarily overdetermined: The Cauchy problem with initial data on the "space-like" line has no solution.

Theorem 2 states that the absence of a core unitary dilation is related to the nonsolvability of the Cauchy problem for an overdetermined system of differential equations.

§2. Functional Models and Hardy Spaces on Riemann Surfaces.

This section is concerned with representations of the semigroup $K \subset R^n$,
$$T(t_1,\ldots,t_n) = \exp i \sum_1^n A_j t_j ,$$
$$\|T(t_1,\ldots,t_n)\| \le 1 ,$$
with infinitessimal operators $A_j : H \to H$ whose imaginary components are finite dimensional:

1) $\|A_j\| < \infty$,

2) $\dim \dfrac{A_j - A_j^*}{2i} H < \infty$.

In general, when $n \ge 3$, the representation $T(t_1,\ldots,t_n)$ admits no unitary dilations. The aim is to construct a unitary dilation of some subrepresentations of finite codimension. Thus the von Neumann inequality
$$\|\varphi(A_1,\ldots,A_n) h\| \le \|h\| \, \|\varphi\|_{H^\infty}$$
will be established for vectors h in some subspace of finite codimension.

The finite dimensional space
$$\operatorname{Im} A_1 H = \operatorname{Im} \Sigma \, a_j A_j H \subset H ,$$
where $(a_1,\ldots,a_n) \in K$, is denoted G. And the orthogonal projection of the space H onto the subspace G is denoted P_G. With every vector h in H is associated the vector function
$$u_h(t_1,\ldots,t_n) = P_G \, T(t_1,\ldots,t_n)^* h$$
with values in the space G.

As is easily seen,
$$\frac{\partial}{\partial t_j} u_k = -i A_j^* u_k ,$$
$$\frac{A_k - A_k^*}{i} \frac{\partial}{\partial t_j} u_k - \frac{A_j - A_j^*}{i} \frac{\partial}{\partial t_k} u_k + (A_k A_j^* - A_j A_k^*) u_k = 0 .$$

On defining
$$\sigma_k = \left. \frac{A_k - A_k^*}{i} \right|_G , \qquad \gamma_{kj} = \left. \frac{A_k A_j^* - A_j A_k^*}{i} \right|_G ,$$
one obtains the hyperbolic system of partial differential equations,
$$\sigma_k \frac{1}{i} \frac{\partial u}{\partial t_j} - \sigma_j \frac{1}{i} \frac{\partial u}{\partial t_k} + \gamma_{jk} u = 0 , \tag{43}$$
$j,k = 1,\ldots,n$, which the vectorfunctions $u_k(t_1,\ldots,t_n)$ satisfy.

Consider the space m_+ of all solutions of the system (43) which have finite energy,
$$\|u\|_{m_+}^2 = \int_0^\infty (\sigma_1 u, u)_G \, dx_1 \ .$$
When n = 2, the conception of a solution in the necessary degree of generality was introduced in Chapter 1. It remains to make use of the fact that each equation of the system (43) contains derivatives in only two of n possible directions.

The truncated shifts
$$u(x_1, \ldots, x_n) \to u(x_1 + t_1, \ldots, x_n + t_n)$$
in m_+ are contractions. The adjoint operators form a strongly continuous contractive representation which is denoted $R_T'(t_1, \ldots, t_n)$ in what follows. When n = 2, the system of equations (43) is not overdetermined. The Cauchy problem is soluble for arbitrary initial data. That means that, when n = 2, the operators R_T' are isometries.

It is not difficult to show that the operator
$$J': h \to u_h(t_1, \ldots, t_n)$$
is a contraction and that
$$(J')^* R_T'(t_1, \ldots, t_n) = T(t_1, \ldots, t_n) (J')^* \ .$$
In other words $(J')^*$ is a contractive morphism of the representations
$$(J')^*: R_T'(t_1, \ldots, t_n) \to T(t_1, \ldots, t_n) \ .$$
The contractions $T(t_1, \ldots, t_n)^*$ generate a semigroup of isometries in the completion with respect to the norm
$$\sqrt{\|h\|^2 - \|J' h\|^2}$$
of the orthogonal complement of Ker J' in H. According to the results of Chapter 1, the semigroup possesses a minimal unitary dilation $U_T(t_1, \ldots, t_n)$ which is unique within unitary equivalence.

As usual in the construction of unitary dilations, the contractive operator J' is extended to an isometric morphism
$$i': T(t_1, \ldots, t_n) \to R_T'(t_1, \ldots, t_n) \oplus U_T(t_1, \ldots, t_n) \ .$$
The orthogonal complement of the range of the operator i' is a common invariant subspace of the operators
$$R_T'(t_1, \ldots, t_n) \oplus U_T(t_1, \ldots, t_n) \ .$$
The respective subrepresentations are denoted $R_T''(t_1, \ldots, t_n)$:
$$i'': R_T''(t_1, \ldots, t_n) \to R_T'(t_1, \ldots, t_n) \oplus U_T(t_1, \ldots, t_n) \ .$$
An exact sequence of representations and morphisms
$$0 \to R_T'' \xrightarrow{i''} R_T' \oplus U_T \xrightarrow{i'*} T \to 0$$
is obtained, in which the operators i" and i' are isometric.

The two-parameter semigroups of contractions $T(t_1,t_2)$ were studied in the previous section. When $n = 2$, the representations $R_T{}''$, $R_T{}' \oplus U_T$ are isometric. In the case $n > 2$, the transition to the exact sequence of morphisms with partially isometric representations succeeds in realizing the "value" of finite dimensions (at the cost of a transition from $R_T{}''$, $R_T{}' \oplus U_T$ to their subrepresentations of finite codimension). Let $I_R{}''$ be a subrepresentation of a representation in the space
$$\{h: \|R_T{}'' h\| = \|h\|\},$$
$I_R{}'$ the subrepresentation of $R_T{}'$ in the space
$$\{h: \|R_T{}' h\| = \|h\|\}.$$
The following is true.

PROPOSITION 1. The codimension of the representation $I_R{}''(t_1,\ldots,t_n)$ in $R_T{}''(t_1,\ldots,t_n)$ and the codimension of the representation $I_R{}'(t_1,\ldots,t_n)$ in $R_T{}'(t_1,\ldots,t_n) \oplus U_T(t_1,\ldots,t_n)$ are finite.

The proof of the proposition is presented in the supplement to Chapter 4.

COROLLARY. The "intertwining" operator j_R:
$$\begin{array}{ccccc}
0 \to I_R{}''(t_1,\ldots,t_n) & \to & I_R{}'(t_1,\ldots,t_n) & \to & T'(t_1,\ldots,t_n) \to 0 \\
\downarrow j_{R''} & & \downarrow j_{R'} & & \downarrow j_R \\
0 \to R_T{}''(t_1,\ldots,t_n) & \to & R_T{}'(t_1,\ldots,t_n) \oplus U(t_1,\ldots,t_n) & \to & T(t_1,\ldots,t_n) \to 0
\end{array}$$
has a range of finite codimension and a kernel of finite dimension.

The ranks of nonunitarity of the isometric representations $I_R{}''(t_1,\ldots,t_n)$, $I_R{}'(t_1,\ldots,t_n)$ are finite. That means that the results of §3 of Chapter 1 are applicable.

The one-dimensional irreducible components of the algebraic set
$$\{(z_1,\ldots,z_n): \bigcap_{k,j=1,\ldots,n} \mathrm{Ker}(\sigma_k z_j - \sigma_j z_k + \gamma_{kj}) \neq \{0\}\}$$
are equipped with the structure of a Riemann surface. As in §3 of Chapter 1, the restrictions of the functions $z_1,\ldots,z_n$ to n-dimensional real space $R^n \subset C^n$ are denoted $\xi_1,\ldots,\xi_n$. According to the results of Chapter 1, finite-dimensional unitary representations $\tau_1{}'',\ldots,\tau_s{}''$ of the fundamental groups of the components $\Gamma_1,\ldots,\Gamma_s$ exist such that the representation $I_R{}''(t_1,\ldots,t_n)$ admits a subrepresentation $I''(t_1,\ldots,t_n)$ of finite codimension which is unitarily equivalent to a representation by multiplication operators
$$f \to \exp i(\Sigma \xi_j t_j) f$$
in the space
$$H^2_{\tau_1{}''}(d\xi_1) \oplus \cdots \oplus H^2_{\tau_s{}''}(d\xi_1).$$

Similarly the representation $I_R'(t_1,\ldots,t_n)$ admits a subrepresentation $I'(t_1,\ldots,t_n)$ of finite codimension which is equal to an orthogonal sum of unitary representations and representations by multiplication operators in the space
$$H^2_{\tau_1}(d\xi_1) \oplus \cdots \oplus H^2_{\tau_s}(d\xi_1) \;.$$
See Chapter 1 for the construction of the Hardy space H^2_τ.

The described representations of finite codimension can be chosen to be embedded in one another. First choose I' and then $I'' \subset I'$.

$$\begin{array}{ccccccc}
0 \to & I''(t_1,\ldots,t_n) & \to & I'(t_1,\ldots,t_n) & \to & T''(t_1,\ldots,t_n) & \to 0 \\
 & \downarrow j'' & & \downarrow j' & & \downarrow j & \\
0 \to & I_R''(t_1,\ldots,t_n) & \to & I_R'(t_1,\ldots,t_n) & \to & T'(t_1,\ldots,t_n) & \to 0 \\
 & \downarrow j_R'' & & \downarrow j_R' & & \downarrow j_R & \\
0 \to & R_T''(t_1,\ldots,t_n) & \to & R_T'(t_1,\ldots,t_n) \oplus U_T & \to & T(t_1,\ldots,t_n) & \to 0 \;.
\end{array}$$

As previously, all morphisms on the rows are isometric or coisometric and j'', j', j_R'', j_R' are isometries.

Thus the operator $J = j_R\, j$ which intertwines the representations $T''(t_1,\ldots,t_n)$ and $T(t_1,\ldots,t_n)$ is isometric on a subspace of finite codimension:
$$\mathrm{codim}\,\{h\colon \|j_R\, j\, h\| = \|h\|\} < \infty, \quad \mathrm{codim}\, j_R\, j\, H < \infty \;.$$

The functional model of the representation T'' can now be obtained by completely standard arguments. The following theorem has been proved.

Let $T(t_1,\ldots,t_n) = \exp i \sum A_j t_j$ be a completely nonunitary contractive representation of the semigroup K and let
$$\|A_j\| < \infty, \quad \dim \mathrm{Im}\, A_j\, H < \infty, \quad j = 1,2,\ldots,n \;.$$

Denote by $\Gamma_{+1}, \Gamma_{+2}, \ldots, \Gamma_{+s}$ the one-dimensional irreducible components of the semialgebraic set
$$\{(z_1,\ldots,z_n)\colon \bigcap_{kj} \mathrm{Ker}(\sigma_k z_j - \sigma_j z_k + \gamma_{kj}) \neq \{0\},\ \mathrm{Im}\, z_1 > 0\} \;.$$
Denote the universal covering spaces by $\hat{\Gamma}_{+1}, \hat{\Gamma}_{+2}, \ldots, \hat{\Gamma}_{+s}$.

THEOREM 3. Finite-dimensional unitary representations $\tau_1', \tau_1'', \tau_2', \tau_2'', \ldots, \tau_s', \tau_s''$ of the fundamental groups of the Riemann surfaces $\Gamma_{+1}, \Gamma_{+2}, \ldots, \Gamma_{+s}$ and contractive operator-functions θ_j, which are holomorphic in $\hat{\Gamma}_{+1}, \hat{\Gamma}_{+2}, \ldots, \hat{\Gamma}_{+s}$ exist such that
$$\theta_j(g\,\lambda) = \tau_j'(g^{-1})\, \theta_j(\lambda)\, \tau_j''(g)$$
and such that the representation $T''(t_1,\ldots,t_n)$, which is equal to the orthogonal sum of representations $T_j''\colon H_j \to H_j$ where H_j is the orthogonal complement of
$$\{\theta\, f \oplus f\colon f \in H^2_{\tau_j''}(d\xi_1)\}$$
in

and
$$H^2_{\tau_j}(d\xi_1) \oplus L^2(\overline{\sqrt{1 - \theta*\theta}} \, d\xi_1) \, ,$$

$$T_j(t_1,\ldots,t_n)(f \oplus g) = P_{H_j}[\exp(i \, \Sigma \, t_j z_j) \, f \oplus \exp(i \, \Sigma \, t_j \xi_j) \, g]$$

admits a Noetherian morphism in the representation $T(t_1,\ldots,t_n)$:

$$J: T''(t_1,\ldots,t_n) \to T(t_1,\ldots,t_n) \, .$$

Here the operator J is isometric on a subspace of finite codimension,
$$\text{codim}\{h: \|J \, h\| = \|h\|\} < \infty \, .$$

The analogue of the von Neumann inequality for multi-parameter semigroups of contractions is obtained with the help of Theorem 3.

Let $A_1,\ldots,A_n$ be pairwise commuting bounded operators in a Hilbert space H. Assume that the inequalities
$$|\text{Im } A_j| \leq C \text{ Im } A_1$$
hold for some positive number C, $j = 2, 3,\ldots,n$.

Consider the algebra $\mathcal{A}$ of finite linear combinations of exponentials
$$\exp(i \, \Sigma \, a_j z_j)$$
satisfying the condition
$$\sum_{2}^{n} |a_j| \leq \frac{1}{C(n-1)} \, a_1 \, .$$

COROLLARY. (ANALOGUE OF THE VON NEUMANN INEQUALITY). Assume that $A_1,\ldots,A_n$ are pairwise commuting bounded operators in H with finite-dimensional imaginary components:
$$\dim(A_j - A_j^*) H < \infty$$
for $j = 1,\ldots,n$. If
$$|\text{Im } A_j| \leq C \text{ Im } A_1$$
for $j = 2, 3,\ldots,n$, then a subspace L of finite codimension in H exists such that the inequality
$$\|\varphi(A_1,\ldots,A_n) \, h\| \leq \sup_{\partial \text{ spset}(A_1,\ldots,A_n)} |\varphi| \, \|h\| \tag{44}$$
holds for every function $\varphi \in \mathcal{A}$ and every vector $h \in L$. Here $\partial \text{ spset}(A_1,\ldots,A_n) \subset R^n$ is the set defined by the formula:
$$\partial \text{ spset}(A_1,\ldots,A_n)$$
$$= \{(\xi_1,\ldots,\xi_n) \in R^n: \exists \, g \in (A_1 - A_1^*) H, \, g \neq 0,$$
$$\forall_{j,h} \, (\frac{A_k - A_k^*}{i} \xi_j - \frac{A_j - A_j^*}{i} \xi_k + \frac{A_k^* A_j - A_j^* A_k}{i}) \, g = 0 \} \, .$$

If the inequality (44) does not hold for some vector $h \in H$, then, as can be seen from the proof of Theorem 3, this vector corresponds to a solution of finite energy of the hyperbolic system in the cone K which does not admit an extension which is a solution of finite energy in the whole space R^n and is equal to zero in $(-K)$. Examples of such nonextendable solutions will be given.

Consider the hyperbolic system

$$\frac{1}{i}\frac{\partial u}{\partial t_1} - \begin{pmatrix} 0 & \sigma_1 \\ \sigma_1 & 0 \end{pmatrix} \frac{1}{i}\frac{\partial u}{\partial x} + \begin{pmatrix} 0 & \sigma_2 \\ \sigma_2 & 0 \end{pmatrix} u = 0$$

$$\frac{1}{i}\frac{\partial u}{\partial t_2} - \begin{pmatrix} 0 & \sigma_2 \\ \sigma_2 & 0 \end{pmatrix} \frac{1}{i}\frac{\partial u}{\partial x} - \begin{pmatrix} 0 & \sigma_1 \\ \sigma_1 & 0 \end{pmatrix} u = 0$$
(45)

where σ_1, σ_2 are Pauli matrices. Look for a solution of the form

$$u(t_1, t_2, x) = v(t_1, t_2) e^{-x} .$$

Then

$$\frac{\partial v}{\partial t_1} = \begin{pmatrix} 0 & \sigma_1 - i\sigma_2 \\ \sigma_1 - i\sigma_2 & 0 \end{pmatrix} v$$

$$\frac{\partial v}{\partial t_2} = \begin{pmatrix} 0 & \sigma_1 + i\sigma_2 \\ \sigma_1 + i\sigma_2 & 0 \end{pmatrix} v .$$

The solutions of the system of equations (45), which are of the form $v(t_1, t_2) e^{-x}$, form, as is easily seen, a four-dimensional common invariant subspace of the operators

$$\frac{1}{i}\frac{\partial}{\partial x}, \quad \frac{1}{i}\frac{\partial}{\partial t_1}, \quad \frac{1}{i}\frac{\partial}{\partial t_2} .$$

Their restrictions to this subspace are denoted A_1, A_2, A_3. It remains to observe that the system (45) does not admit solutions of finite energy in R^3 and that the set $\mathrm{spset}(A_1, A_2, A_3)$ consists of the single point $\{0, 0, i\}$ since the system of equations

$$z_1 h = \left[\begin{pmatrix} 0 & \sigma_1 \\ \sigma_1 & 0 \end{pmatrix} z_3 - \begin{pmatrix} 0 & \sigma_2 \\ \sigma_2 & 0 \end{pmatrix} \right] h \qquad (46)$$

$$z_2 h = \left[\begin{pmatrix} 0 & \sigma_2 \\ \sigma_2 & 0 \end{pmatrix} z_2 - \begin{pmatrix} 0 & \sigma_1 \\ \sigma_1 & 0 \end{pmatrix} \right] h \qquad (47)$$

implies the equation

$$\begin{pmatrix} [\sigma_1, \sigma_2] & 0 \\ 0 & [\sigma_1, \sigma_2] \end{pmatrix} (z_3^2 + 1) h = 0 .$$

Wherefrom it follows that $z_3 = +i$ in view of the inequality $\mathrm{Im}\, z_3 \geq 0$, and z_1, z_2 are obtained from equations (46), (47).

REMARK 2. Because of Proposition 1 the representation $T(t_1, \ldots, t_n)^*$ has a subrepresentation of finite codimension which admits an isometric morphism into a coisometric representation

$$[I_R{}'(t_1,\ldots,t_n) \oplus U_R(t_1,\ldots,t_n)]* \ .$$

That means that $T(t_1,\ldots,t_n)*$ has a subrepresentation of finite codimension which admits a unitary dilation. The following result is obtained when the representation $T(t_1,\ldots,t_n)*$ is replaced by the representation $T(t_1,\ldots,t_n)$ in all the preceding constructions.

PROPOSITION 2. If
$$T(t_1,\ldots,t_n) = \exp i \ \Sigma \ A_j t_j$$
is a contractive representation of the semigroup K such that
$$\|A_j\| < \infty, \quad \dim(A_j - A_j^*) H < \infty, \quad j = 1, 2, \ldots, n,$$
then the restrictions of the operators $T(t_1,\ldots,t_n)$ to some common invariant subspace of finite codimension form a representation which admits a unitary dilation.

SUPPLEMENT 1. On Overdetermined Hyperbolic Systems of Equations with Constant Coefficients.

Consider the hyperbolic system of equations
$$\frac{1}{i} \frac{\partial u}{\partial t_j} = \frac{1}{i} \sigma_j \frac{\partial u}{\partial x} + \gamma_j \ u \tag{48}$$
for $j = 1, 2, \ldots, n-1$ where σ_j, γ_j are selfadjoint operators in a finite-dimensional Hilbert space E.

The Cauchy problem for the system (48) in R^n consists of the search for smooth solutions of this system which coincide for $t_1 = t_2 = \cdots = t_{n-1} = 0$ with smooth functions $f(x)$ such that
$$\int_{-\infty}^{+\infty} \|f(x)\|_E^2 \ dx < \infty \ .$$

It is well-known that a solution of the Cauchy problem is unique, but that it does not exist for all initial data $f(x)$ in the case $n > 2$. A condition which is sufficient for a solution of the Cauchy problem is needed. Pass to Fourier transforms of solutions considered as functions of the variable x,
$$f(x) \to \hat{f}(\xi) = \frac{1}{\sqrt{2\pi}} \int_{-\infty}^{+\infty} f(x) \ e^{-ix\xi} \ dx \ ,$$
$$u(x,t_1,\ldots,t_n) \to \hat{u}(\xi,t_1,\ldots,t_n) = \frac{1}{\sqrt{2\pi}} \int_{-\infty}^{+\infty} u \ e^{-ix\xi} \ dx \ .$$

It follows that
$$\frac{1}{i} \frac{\partial \hat{u}}{\partial t_j} = (\sigma_j \xi + \gamma_j) \ \hat{u}$$
where
$$\hat{u}|t_1 = \cdots = t_n = 0 = \hat{f} \ .$$

A sufficient condition for the existence of a solution of this system is that the vector
$$(\sigma_{j_1}\xi + \gamma_{j_1})(\sigma_{j_2}\xi + \gamma_{j_2})\cdots(\sigma_{j_n}\xi + \gamma_{j_n})\hat{f}(\xi) \in E$$
is independent of the order of multiplication for almost ξ. The solution is of the form
$$\hat{u}(t_1,\ldots,t_{n-1},\xi) = \exp i \Sigma\, t_j(\sigma_j \xi + \gamma_j)\hat{f}(\xi). \tag{49}$$
In other words, a sufficient condition for the existence of a solution is that the identity
$$[(\sigma_{j_1}\xi + \gamma_{j_1}),(\sigma_{j_2}+\gamma_{j_2})](\sigma_{j_3}\xi + \gamma_{j_3})\cdots(\sigma_{j_m}\xi + \gamma_{j_m})\hat{f}(\xi) = 0$$
holds for almost all ξ for all finite sequences of natural numbers $j_1, j_2, \ldots, j_m$.

Enumerate all finite sequences of natural numbers which do not exceed $n-1$: $k \to j_{1k}, j_{2k}, \ldots$. The set Γ_r of vectors (f,ξ) of the complex vector space $E \oplus C^1$, which satisfy the equations
$$[(\sigma_{j_{1k}}\xi + \gamma_{j_{1k}}),(\sigma_{j_{2k}}\xi + \gamma_{j_{2k}})](\sigma_{j_{3k}}\xi + \gamma_{j_{3k}})\cdots f = 0$$
for $k = 1, 2, \ldots, r$, is an algebraic set and
$$\Gamma_1 \supset \Gamma_2 \supset \Gamma_3 \supset \cdots .$$
That means that the sequence Γ_r stabilizes:
$$\Gamma_{r_0} = \Gamma_{r_0+1} = \Gamma_{r_0+2} = \cdots .$$
So a sufficient condition for the existence of a solution of the Cauchy problem with initial data $f(\xi)$ is that the condition
$$[(\sigma_{j_{1k}}\xi + \gamma_{j_{1k}}),(\sigma_{j_{2k}}\xi + \gamma_{j_{2k}})](\sigma_{j_{3k}}\xi + \gamma_{j_{3k}})\cdots \hat{f}(\xi) = 0$$
holds for almost all ξ for $1 \leq k \leq r_0$. The following result is obtained on applying the inverse Fourier transformation in the variable ξ.

PROPOSITION 1. Natural numbers N and r_0 exist such that for an arbitrary N-times differentiable vector-function $f(x)$ on the line with values in the space E, which satisfies the inequality
$$\int_{-\infty}^{+\infty} \|f\|_E^2 \, dx < \infty$$
and the identities
$$[(\sigma_{j_{1k}}\tfrac{1}{i}\tfrac{d}{dx} + \gamma_{j_{1k}}),(\sigma_{j_{2k}}\tfrac{1}{i}\tfrac{d}{dx} + \gamma_{j_{2k}})](\sigma_{j_{3k}}\tfrac{1}{i}\tfrac{d}{dx} + \gamma_{j_{3k}})\cdots f = 0$$
when $k = 1, 2, \ldots, r_0$, the formula
$$u = \exp i \Sigma\, t_j(\sigma_j \tfrac{1}{i}\tfrac{d}{dx} + \gamma_j)\, f(x)$$
defines a solution of the Cauchy problem
$$\tfrac{1}{i}\tfrac{\partial u}{\partial t_j} = \sigma_j \tfrac{1}{i}\tfrac{\partial u}{\partial x} + \gamma_j u$$

with the initial data
$$u|_{t_1 = t_2 = \cdots = t_{n-1} = 0} = f(x).$$

COROLLARY 1. Let L be the space of all solutions in the cone K of the system (48) which are of finite energy and which are N-times continuously differentiable in $\overline{K}$. The function $u \in L$ which are extendable to solutions of finite energy in R^n and which are equal to zero in the cone $-K$ form a vector space of finite codimension in L.

PROOF. Let $f(x)$ be the restriction to the half-line $t_1 = 0, \ldots, t_{n-1} = 0$, $x > 0$, of a solution belonging to L. Extend the function to be zero on the left half-line. It is clear that, by restricting oneself to a vector subspace of finite codimension in L, one can assume that the extended function is N-times differentiable at the origin, and hence on the whole line. It remains to apply the previous proposition to the extended function and to obtain the identity
$$u|_{-K} \equiv 0$$
from the equation $f(x) = 0$ when $x < 0$ and from (49). As in the previous chapter, it is assumed that
$$x + \sum_{1}^{n-1} t_j \sigma_j \geq 0$$
when $(x, t_1, \ldots, t_n) \in K$.

REMARK 1. The assertion made in Proposition 1, Section 2 of Chapter 4, is a consequence of the above proved Corollary 1.

APPENDIX

TRIANGULAR MODELS OF COMMUTING LINEAR OPERATORS

AND NONLINEAR DIFFERENTIAL EQUATIONS

Consider a pair A_1, A_2 of commuting continuous linear operators in a Hilbert space H. The operator A_1 is assumed to be a Volterra (a completely continuous operator with no nonzero spectrum) which is also dissipative and whose imaginary component is finite-dimensional. The "boundary" space

$$\frac{A_1 - A_1^*}{2i} H$$

is denoted G, and its orthogonal projection is denoted P_G.

It is well-known that the operators A_1, A_2 possess nontrivial common invariant subspaces, for example

$$\text{Ker} \exp(-it\, A_1^{-1}) \neq H$$

for small t. Moreover a maximal nested chain of orthogonal projections P_t, $0 \leq t \leq 1$, onto the common invariant subspaces for both operators exists. This allows one to obtain a triangular model for the pair A_1, A_2. The discussion is restricted to the case

$$(A_2 - A_2^*) H \subset (A_1 - A_1^*) H$$

and it is assumed that the operator-function P_t is continuously differentiable in the interval (0,1) and that

$$\frac{d}{dt}(P_G\, P_t|_G) > 0 .$$

Then according to a well-known result of M. S. Livšic, a unitary linear transformation exists of H into $L^2(\pi(t), dt)$, $\pi(t) \geq 0$, $0 \leq t \leq 1$, where the operator A_1 corresponds to the operator

$$\hat{A}_1 : f(x) \to i \int_x^1 \pi(t) f(t)\, dt \qquad (1)$$

and the subspace $\mathcal{L}_t = P_t H$ corresponds to

$$\chi_{(0,t)}\, L^2(\pi(t), dt) .$$

In the work of M. S. Livšic, the operator $\hat{A}_1$ is defined by the formula

$$f \to i\, \Pi(x) \int_x^1 \Pi(t) f(t)\, dt .$$

Replace f by $\Pi(t)^{-1} f = g$ and define $\pi = \Pi^2$ so that

$$\int \|f\|^2\, dt = \int (\pi(t) g, g)\, dt .$$

The operators $\pi(t): G \to G$ are equal to
$$\frac{d}{dt} |A_1 - A_1^*|^{\frac{1}{2}} P_t |A_1 - A_1^*|^{\frac{1}{2}} \Big|_G ,$$
and the unitary operator U such that
$$U^{-1} \hat{A} U = A$$
is defined on a vector $g \in G$ by the formula
$$(U g)(t) = |A_1 - A_1^*|^{-\frac{1}{2}} g .$$

PROPOSITION 1. The pair of operators A_1, A_2 is unitarily equivalent to the pair of operators
$$\hat{A}_1 : f(t) \to i \int_t^1 \pi(t) f(t) dt ,$$
$$\hat{A}_2 : f(t) \to \gamma(t) f(t) + i \int_t^1 \sigma \pi(\tau) f(\tau) d\tau \qquad (2)$$
in the space $L^2(\pi(t) dt)$. Here
$$\sigma = |A_1 - A_1^*|^{-\frac{1}{2}} \frac{A_2 - A_2^*}{i} |A_1 - A_1^*|^{-\frac{1}{2}} \qquad (3)$$
and the selfadjoint operator-function $\gamma(t)$ is defined by the differential equation
$$i \frac{d\gamma}{dt} = [\pi(t), \sigma] \qquad (4)$$
and the "initial" condition
$$\gamma(1) = |A_1 - A_1^*|^{-\frac{1}{2}} \frac{A_1 A_2^* - A_2 A_1^*}{i} |A_1 - A_1^*|^{-\frac{1}{2}} . \qquad (5)$$

PROOF. Since
$$\frac{A_1 - A_1^*}{i} f_0 = \int_0^1 \pi(t) f(t) dt ,$$
if $f(t) \equiv f_0$, then
$$U \frac{A_2 - A_2^*}{i} U^{-1} f = |A_1 - A_1^*|^{-\frac{1}{2}} \frac{A_2 - A_2^*}{i} |A_1 - A_1^*|^{-\frac{1}{2}} f_0$$
$$= |A_1 - A_1^*|^{-\frac{1}{2}} \frac{A_2 - A_2^*}{i} |A_1 - A_1^*|^{-\frac{1}{2}} \int_0^1 \pi(\tau) f(\tau) d\tau$$
$$= \int_0^1 \sigma \pi(\tau) f(\tau) d\tau .$$
That means that the operator
$$U A_2 U^{-1} - i \int_x^1 \sigma \pi(t) dt$$
is selfadjoint, and the spaces
$$\chi_{(0,t)} L^2(\pi(t) dt)$$
are invariant subspaces of the operator.

Therefore the identity
$$U A_2 U^{-1} f = \gamma f + i \int_t^1 \sigma \pi(\tau) f(\tau) d\tau$$
holds for some bounded measurable operator-function $\gamma(\tau)$ for which
$$\gamma^*(t) \pi(t) = \pi(t) \gamma(t) . \qquad (6)$$
It remains to verify equations (4) and (5). Use is made of the identities
$$\hat{A}_1 \hat{A}_2 f = i \int_t^1 \pi(\tau) \gamma(\tau) f(\tau) d\tau$$
$$- \int_t^1 \pi(\tau) \sigma \int_\tau^1 \pi(\tau') f(\tau') d\tau' d\tau ,$$
$$\hat{A}_2 \hat{A}_1 f = i \gamma(t) \int_t^1 \pi(\tau) f(\tau) d\tau$$
$$= \int_t^1 \sigma \pi(\tau) \int_\tau^1 \pi(\tau') f(\tau') d\tau'$$

On defining
$$g = \int_\tau^1 \pi(\tau') f(\tau') d\tau') ,$$
one obtains
$$0 = (\hat{A}_1 \hat{A}_2 - \hat{A}_2 \hat{A}_1) f$$
$$= i \int_t^1 \pi(\tau) \gamma(\tau) f(\tau) d\tau - i \gamma(t) \int_t^1 \pi(\tau) f(\tau) d\tau$$
$$- \int_t^1 [\pi(\tau),\sigma] g(\tau) d\tau$$
$$= -i \int_t^1 [\gamma(t),\pi(\tau)] f(\tau) d\tau$$
$$- \int_t^1 ([\pi(\tau),\sigma] - i \frac{d\gamma}{d\tau}) g d\tau .$$

That means that
$$0 = i [\gamma(t),\pi(t)] \pi(t)^{-1} \frac{dg}{dt} + i ([\pi,\tau] - i \frac{d\gamma}{dt}) g .$$
On varying independently the values of the function g and its derivative at the point $t \in (0,1)$, one obtains the equations
$$[\gamma(t),\pi(t)] = 0 ,$$
$$i \frac{d\gamma}{dt} = [\pi(t),\sigma] . \qquad (7)$$
Thus because of (6) and (7), $\gamma = \gamma^*$.

It remains to derive equation (5). Since
$$\frac{A_1 - A_1^*}{i} f = \int_0^1 \pi(t) f(t) dt = i A_1^* f|_{t=1} ,$$
$$\frac{\hat{A}_2 - \hat{A}_2^*}{i} f = \sigma \int_0^1 \pi(t) f(t) dt = i A_2^* f|_{t=1} - i \gamma(1) f|_{t=1}$$

and in view of the commutativity of the operators A_1^*, A_2^*, one obtains

$$\left(\frac{\hat{A}_1 - \hat{A}_1^*}{i} \hat{A}_2^* - \frac{\hat{A}_2 - \hat{A}_2^*}{i} \hat{A}_1^* \right) f = i \gamma(1) A_1^* f \big|_{x=1}$$

$$= \gamma(1) \int_0^1 \pi(\tau) f(\tau) d\tau .$$

From this, on defining $f = U g$, $g \in G$, one obtains

$$|A_1 - A_1^*|^{-\frac{1}{2}} \left(\frac{A_1 - A_1^*}{i} A_2^* - \frac{A_2 - A_2^*}{i} A_1^* \right) g$$

$$= \gamma(1) \int_0^1 \pi(\tau) d\tau \, |A_1 - A_1^*|^{-\frac{1}{2}} g .$$

And finally

$$|A_1 - A_1^*|^{-\frac{1}{2}} \left(\frac{A_1 - A_1^*}{i} A_2^* - \frac{A_2 - A_2^*}{i} A_1^* \right) |A_1 - A_1^*|^{-\frac{1}{2}} = \gamma(1) .$$

This completes the proof of Proposition 1.

COROLLARY 1. Assume that $\sigma = \sigma^*$ and that $\gamma(t) = \gamma(t)^*$. The operators $\hat{A}_1$ and $\hat{A}_2$ commute when, and only when,

$$[\gamma(t), \pi(t)] = 0$$

$$i \frac{d\gamma}{dt} = [\pi(t), \sigma] .$$

Assume that σ, $\gamma = \gamma(1)$ are two selfadjoint operators in a finite-dimensional Hilbert space. What are the conditions on a monotone increasing smooth operator-function $X(t)$,

$$\pi = \frac{dX}{dt} < \infty ,$$

that the operators $\hat{A}_1$, $\hat{A}_2$ with

$$\gamma(t) = \gamma + i \int_t^1 [\pi(\tau), \sigma] d\tau$$

commute? The following holds because of Corollary 1.

PROPOSITION 2. The operators $\hat{A}_1$ and $\hat{A}_2$ commute when, and only when, the smooth operator-function $X(t)$, with $X(1) = 0$ and

$$\frac{dX}{dt} < 0,$$

is a solution of the following nonlinear differential equation with constant coefficients:

$$\left[\frac{dX}{dt} , \gamma + [X(t), \sigma] \right] = 0 . \tag{8}$$

Thus the construction of a pair of commuting operators reduces to the search for solutions of the nonlinear equation (8). The discussion will be restricted to indicating the choice of integrals of equations (8). Because of the identities

$$[\gamma, \pi] = 0, \quad i \frac{d\gamma}{dt} = [\pi \cdot \sigma], \quad \pi = \pi^* ,$$

the operators $\lambda \sigma + \gamma$ and $\lambda \sigma + \gamma(t)$ are unitarily equivalent for all real λ. That means that the following is true.

PROPOSITION 3. Let $X(t)$ be a solution of equation (8) and let
$$\gamma(t) = \gamma + [X(t), \sigma].$$
Then the coefficients of the polynomial
$$d(z_1, z_2) = \det(\sigma z_1 - z_1 + \gamma(t))$$
are independent of t.

REMARK. From formulas (3) and (5), it follows that the algebraic curve $\{(z_1, z_2): d(z_1, z_2) = 0\}$ is a canonical spectral set for the pair of operators
$$\hat{A}_1|_{\chi(1-t,1)} L^2(\pi\, dt), \quad \hat{A}_2|_{\chi(1-t,1)} L^2(\pi\, dt).$$

In conclusion an example is considered which is related to operators with three-dimensional imaginary components.

Choose three distinct nonzero real numbers $\alpha_1, \alpha_2, \alpha_3$ whose sum is equal to zero.

The triangular model will be obtained for a pair of commuting bounded operators A_1, A_2, with three-dimensional imaginary components, whose canonical spectral set is
$$\Gamma_+ = \{(z_1, z_2): (z_2 - \alpha_1)(z_2 - \alpha_2)(z_2 - \alpha_3) - z_1^2 z_2 = 0, \text{ Im } z_1 > 0\}$$
and whose common spectrum consists of the three points
$$(0, \alpha_1)\ (0, \alpha_2),\ (0, \alpha_3).$$

As was shown in Chapter 2, such a pair of operators can be obtained as the restriction of the operators
$$L_1 = -i\, \tfrac{d}{dx}, \quad L_2 = -i\, \sigma\, \tfrac{d}{dx} + \gamma$$
where
$$\sigma = \begin{pmatrix} -1 & 0 & 0 \\ 0 & 0 & 0 \\ 0 & 0 & 1 \end{pmatrix}, \quad \gamma = \begin{pmatrix} 0 & a_1 & a_2 \\ a_1 & 0 & a_1 \\ \overline{a_2} & a_1 & 0 \end{pmatrix},$$
$$2(\text{Re } a_2)\, a_1^2 = \alpha_1 \alpha_2 \alpha_3,$$
$$2 a_1^2 + |a_2|^2 = -\alpha_1 \alpha_3 - \alpha_2 \alpha_3 - \alpha_1 \alpha_3,$$
(9)

to the orthogonal complement of one of their common invariant subspaces (with the corresponding projection onto this subspace). Thus the triangular model of the pair A_1, A_2 is defined by a continuous one-parameter family $\mathcal{L}_t$, $t \in (0,1)$, of common invariant subspaces of the operators L_1, L_2.

The characteristic operator-functions corresponding to these invariant subspaces and defining them by the Beurling-Lax formula are of the form (see Chapter 2)

$$e(z,t) = \sum_{\{\lambda: z_1(\lambda) = z\}} \varphi(\lambda,t) \begin{pmatrix} \psi_1(a_1(t),a_2(t),\lambda) \\ \psi_2(a_1(t),a_2(t),\lambda) \\ \psi_3(a_1(t),a_2(t),\lambda) \end{pmatrix}$$

$$\times (\psi_1(a_1,a_2,\overline{\lambda}), \psi_2(a_1,a_2,\overline{\lambda}), \psi_3(a_1,a_2,\overline{\lambda}))$$

where $\varphi(\lambda,t)$ is a holomorphic $e^{i\tau(a_1(t),a_2(t)) - i\tau(a_1,a_2)}$-periodic function in the strip. From the condition on the common spectrum of the operators A_1, A_2, it follows that the only poles of the differential $d \ln \varphi$ are the three points of the strip for which

$$\begin{cases} z_1(\lambda_1) = 0 \\ z_2(\lambda_1) = \alpha_1 \end{cases} \quad \begin{cases} z_1(\lambda_2) = 0 \\ z_2(\lambda_2) = \alpha_2 \end{cases} \quad \begin{cases} z_1(\lambda_3) = 0 \\ z_2(\lambda_3) = \alpha_3 \end{cases}$$

These poles are simple since the inequality
$$\varphi(\lambda,t) < 1$$
holds at interior points of the strip and
$$|\varphi| = 1$$
on the boundary. When $0 < \text{Im } \lambda < \omega_2$,
$$\varphi(\lambda + \omega_1, t) = e^{i\tau(a_1(t),a_2(t)) - i\tau(a_1,a_2)} \varphi(\lambda)$$

That means that
$$\varphi(\lambda,t) = \text{const} \times \exp(-i \sum_1^3 x_j(t) \, \theta(\omega_1,\omega_2,\lambda - \lambda_j)) ,$$
where $x_j(t) > 0$ and θ is the Weierstrass θ-function.

From the results of Chapter 2, it follows that the invariant subspaces are uniquely determined by the function $\varphi(\lambda,t)$, whereby the choice of multiplicative constant is inessential. The invariant subspaces are monotone decreasing if, and only if, $x_1(t)$, $x_2(x)$, $x_3(t)$ are monotone increasing. In this way is established a one-to-one correspondence of the set of pairs $\hat{A}_1, \hat{A}_2$, defined by formulas (1) and (2) and belonging to the classes studied, and the sets
$$(a_1, a_2, x_1(t), x_2(t), x_3(t))$$
where $t \in (0,1)$, a_1 and a_2 are related by (9), and the functions $x_j(t)$ are continuous, monotone, and bounded with
$$\lim_{t \to 0} x_j(t) = 0 .$$

If one knows $x_1(t), x_2(t), x_3(t)$, one can eliminate $\pi(t)$ from the system of equations (4). By making use of the commutativity of the operators $\pi(t)$ and $\gamma(t)$ and the fact that the spectrum of the operator $\gamma(t)$ coincides for all t with the set $\{\alpha_1, \alpha_2, \alpha_3\}$, it is not difficult to obtain the identity

$$\pi(t) = x_1 \frac{(\gamma-\alpha_2)(\gamma-\alpha_3)}{(\alpha_1-\alpha_2)(\alpha_1-\alpha_3)} + x_2 \frac{(\gamma-\alpha_1)(\gamma-\alpha_3)}{(\alpha_2-\alpha_1)(\alpha_2-\alpha_3)} + x_3 \frac{(\gamma-\alpha_1)(\gamma-\alpha_3)}{(\alpha_3-\alpha_1)(\alpha_3-\alpha_2)} .$$

One arrives at the nonlinear equation

$$i \frac{d\gamma}{dt} = [\; x_1(t)\frac{(\gamma(t)-\alpha_2)(\gamma(t)-\alpha_3)}{(\alpha_1-\alpha_2)(\alpha_1-\alpha_3)}$$
$$+ x_2(t)\frac{(\gamma(t)-\alpha_1)(\gamma(t)-\alpha_3)}{(\alpha_2-\alpha_1)(\alpha_2-\alpha_3)} \qquad (10)$$
$$+ x_3(t)\frac{(\gamma(t)-\alpha_1)(\gamma(t)-\alpha_2)}{(\alpha_3-\alpha_1)(\alpha_3-\alpha_2)} ,\; \sigma]$$

and the initial condition

$$\gamma(1) = \begin{pmatrix} 0 & a_1 & a_2 \\ a_1 & 0 & a_1 \\ \overline{a}_2 & a_1 & 0 \end{pmatrix} . \qquad (11)$$

In this way is obtained a description within unitary equivalence of the set of all pairs of operators A_1, A_2 belonging to the class considered with a given chain of common invariant subspaces. In particular, for every pair considered, a description is given of all maximal chains of projections onto their common invariant subspaces. The set is isomorphic to the set of all those curves which connect the ends of the principal diagonal of a cube in such a way that the tangents of the angles of inclination of the tangent lines with the coordinate axes are positive. Among all chains of invariant subspaces, select the one for which $x_1(t), x_2(t), x_2(t)$ are constants.

PROPOSITION 4. If a_1 and a_2 satisfy relations (9) and x_1, x_2, x_3 are positive numbers, then formulas (1) and (2) where
$$\sigma = \text{diag}(+1, 0, -1)$$
and $\gamma(t)$ is a solution of equation (10) with constant coefficients which satisfies the "initial" condition (11), define a pair of commuting bounded operators with three-dimensional imaginary component with canonical spectral set
$$\{(z_1, z_2) : (z_2-\alpha_1)(z_2-\alpha_2)(z_2-\alpha_3) - z_1^2 z_2 = 0,\; \text{Im } z_1 > 0\}$$
and with common spectrum
$$\{(0,\alpha_1),\; (0,\alpha_2),\; (0,\alpha_3)\} .$$
Among the pairs of operators
$$\hat{A}_1(a_1, a_2, x_1, x_2, x_3),\quad \hat{A}_2(a_1, a_2, x_1, x_2, x_3)$$
constructed, no two are unitarily equivalent.

THEOREM 1. Every pair of commuting operators with three-dimensional imaginary component, whose canonical spectral set is
$$\{(z_1,z_2): (z_2-\alpha_1)(z_2-\alpha_2)(z_2-\alpha_3) - z_1^2 z_2 = 0\}$$
and whose common spectrum is
$$\{(0,\alpha_1), (0,\alpha_2), (0,\alpha_3)\}$$
is unitarily equivalent to one of the above constructed pairs
$$\hat{A}_1(a_1,a_2,x_1,x_2,x_3), \quad \hat{A}_2(a_1,a_2,x_1,x_2,x_3)$$
of the four-parameter family.

Note that the set of solutions (a_1,a_2) of the system of equations (9) is topologically equivalent to a circle, as was shown in Chapter 2.

According to a well-known result of M. S. Livšic, every dissipative Volterra operator with one-dimensional imaginary component is unitarily equivalent to the operator
$$A(\ell): f \to i \int_x^\ell f(t)\,dt$$
in $L^2(0,\ell)$.

Theorem 1 can be seen as an analogue of this theorem of M. S. Livšic.

In conclusion it will be shown that the nonlinear differential (10) with the initial condition (11) can be integrated by quadratures in the case of constant coefficients. The operators $\lambda \sigma + \gamma(t)$ are pairwise unitarily equivalent for every real number λ. In this case, as was shown in Chapter 2, the line bundle $\lambda \sigma + \gamma(t)$ is unitarily equivalent to a bundle of the form

$$\lambda \sigma + \begin{pmatrix} 0 & a_1(t) & a_2(t) \\ \overline{a_1(t)} & 0 & a_1(t) \\ \overline{a_2(t)} & \overline{a_1(t)} & 0 \end{pmatrix}$$

where
$$2\,a_1(t)^2 \operatorname{Re} a_2(t) = \alpha_1 \alpha_2 \alpha_3$$
$$2\,a_1(t)^2 + |a_2(t)|^2 = -\alpha_1\alpha_2 - \alpha_2\alpha_3 - \alpha_1\alpha_3 .$$
(12)

In other words the matrix $\gamma(t)$ can be found in form

$$\begin{pmatrix} \delta(t) & 0 & 0 \\ 0 & 1 & 0 \\ 0 & 0 & \delta(t)^{-1} \end{pmatrix} \begin{pmatrix} 0 & a_1(t) & a_2(t) \\ \overline{a_1(t)} & 0 & a_1(t) \\ \overline{a_2(t)} & \overline{a_1(t)} & 0 \end{pmatrix} \begin{pmatrix} \delta(t)^{-1} & 0 & 0 \\ 0 & 1 & 0 \\ 0 & 0 & \delta(t) \end{pmatrix}.$$

The special form of the diagonal unitary matrix is related to its commutativity with σ and to the symmetry of the matrix $\gamma(t)$ with respect to the cross diagonal. This symmetry is a consequence of the symmetry of equation (10) and the initial condition (11).

As was shown in Chapter 2, the set of solutions (a_1, a_2) of the system (12) is topologically equivalent to a circle. The inverse function $\tau(a_1, a_2)$, which was introduced in Chapter 2, gives a natural parametrization

$$\tau \to (a_1(\tau), a_2(\tau))$$

of this circle.

On observing that

$$\tau(t) = (x_1 + x_2 + x_3)t + \tau_0,$$

and hence that

$$\frac{d\tau}{dt} = x_1 + x_2 + x_3,$$

one arrives at the differential equation

$$i(x_1 + x_2 + x_3)\frac{d\gamma(\tau)}{d\tau}$$

$$= [x_1 \frac{(\gamma(\tau) - \alpha_2)(\gamma(\tau) - \alpha_3)}{(\alpha_1 - \alpha_2)(\alpha_1 - \alpha_3)}$$

$$+ x_2 \frac{(\gamma(\tau) - \alpha_1)(\gamma(\tau) - \alpha_3)}{(\alpha_2 - \alpha_1)(\alpha_2 - \alpha_3)}$$

$$+ x_3 \frac{(\gamma(\tau) - \alpha_1)(\gamma(\tau) - \alpha_2)}{(\alpha_3 - \alpha_1)(\alpha_3 - \alpha_2)}, \sigma]$$

where

$$\gamma(\tau) = \begin{pmatrix} \delta(\tau) & 0 & 0 \\ 0 & 1 & 0 \\ 0 & 0 & \delta(\tau)^{-1} \end{pmatrix} \begin{pmatrix} 0 & a_1(\tau) & a_2(\tau) \\ a_1(\tau) & 0 & a_1(\tau) \\ \overline{a_2}(\tau) & a_1(\tau) & 0 \end{pmatrix} \begin{pmatrix} \delta(\tau)^{-1} & 0 & 0 \\ 0 & 1 & 0 \\ 0 & 0 & \delta(\tau) \end{pmatrix}.$$

On substituting the expression for $\gamma(\tau)$ in the differential equation and cancelling the left and right diagonal factors

$$\begin{pmatrix} \delta(\tau) & 0 & 0 \\ 0 & 1 & 0 \\ 0 & 0 & \delta(\tau)^{-1} \end{pmatrix}, \begin{pmatrix} \delta(\tau)^{-1} & 0 & 0 \\ 0 & 1 & 0 \\ 0 & 0 & \delta(\tau) \end{pmatrix},$$

one arrives at a linear first-order equation for the determination of the scalar function

$$\frac{d}{d\tau} \ln \delta.$$

REFERENCES

1. B. Sz. Nagy and C. Foias, "Harmonic Analysis of Operators in Hilbert Space," North-Holland, Amsterdam, 1970.

2. I. Suciu, "Algebre de Functii," Edetura Academici Republicii Socialisti România, Bucarest, 1969.

3. W. Rudin, "Function Theory in Polydisks," Benjamin, New York, 1969.

4. M. S. Brodskij, "Triangular and Jordan Representations of Linear Operators," American Mathematical Society Translations, vol. 32, Providence, 1971.

5. M. Hasumi, Invariant subspace theorems for Riemann spaces, in "Function Algebras," Scott, Foremann & Co., Chicago, 1965, pp. 250-256.

6. M. Voichick, Invariant subspaces on Riemann surfaces, Canadian J. Math. 18 (1966), 399-403.

7. M. I. Knopp, A corona theorem for automorphic forms and related results, Amer. J. Math. 91 (1969), 599-618.

8. N. L. Alling, A proof of the Corona conjecture for finite open Riemann surfaces, Bull. Amer. Math. Soc. 70 (1964), 110-112.

9. C. M. Stanton, The closed ideals in a function algebra, Trans. Amer. Math. Soc. 154 (1971), 289-300.

10. E. L. Stout, Bounded holomorphic functions on finite Riemann surfaces, Trans. Amer. Math. Soc. 120 (1965), 255-285.

11. M. B. Abrahamse and R. G. Douglas, A class of subnormal operators related to multi-connected domains, Advances in Math. 19 (1976), 106-148.

12. P. D. Lax and R. S. Phillips, "Scattering Theory," Academic Press, New York, 1967.

13. T. Kato, "Perturbation Theory of Linear Operators," Springer-Verlag, Heidelberg, 1966.

14. N. J. Ahiezer, "Elements of the Theory of Elliptic Functions," Nauka, Moscow, 1970). (Russian)

15. B. V. Limaye, Blaschke products for finite Riemann surfaces, Studia Math. 34 (1970), 169-176.

16. L. L. Waksman, Harmonic analysis of multi-parameter semigroups of contractions, Viniti 3991-80, Dep. 1-166. (Russian)

17. I. G. Petrovskij, "Lectures on Partial Differential Equations," Wiley, New York, 1954.

18. N. K. Nikolskij, "Lectures on Shift Operators," Springer-Verlag, to appear.

19. P. A. Fuhrmann, On the Corona theorem and its applications to spectral problems in Hilbert space, Trans. Amer. Math. Soc. 1932 (1968), 55-66.

20. M. S. Livšic, Cayley-Hamilton theorem, vector bundles and divisors of commuting operators, Integral Equations and Operator Theory 6 (1983), 250-273.

MIX
Papier aus verantwortungsvollen Quellen
Paper from responsible sources
FSC® C105338

If you have any concerns about our products,
you can contact us on
ProductSafety@springernature.com

In case Publisher is established outside the EU,
the EU authorized representative is:
**Springer Nature Customer Service Center GmbH
Europaplatz 3, 69115 Heidelberg, Germany**

Printed by Libri Plureos GmbH
in Hamburg, Germany